To Earline
with warmest
wishes
Bob Bowman

The Mystery Of Lady Bountiful

Bob Bowman
Doris Bowman
Edward J. Barrett

ISBN # 1-878096-68-0

Second Printing, 2003

Published by Best of East Texas Publishers
515 South First Street/PO Box 1647
Lufkin, Texas 75902
Telephone 936/634-7444/ FAX 936/634-7750

Dedication

To Hiram Arnold, MD, and George Henderson, Jr, for their encouragement and friendship.

Introduction

In the pre-dawn hours of November 22, 1922, a shot broke the silence of lumber millionaire Hiram Knox's home near Hemphill, a rural sawmill community in easternmost Texas.

When his family entered his bedroom, they found him sprawled across the bed, a bullet in his head and a .45-caliber pistol in his right hand.

It appeared to be a simple case of suicide, but as Sheriff George Alford probed deeper into the death, Knox's charismatic wife Lillian—widely known as Lady Bountiful for her lavish spending, generosity and charitable endeavors—emerged as a murder suspect.

Alford charged Lady Bountiful with murder on his last day in office, igniting a firestorm of newspaper headlines, a wave of speculation within the American lumber industry, and a controversy still debated in Texas after eighty years.

Lost in the 1922 controversy is Lillian Knox's remarkable life, before and after the death of her husband.

Born into a poor Comanche County family in 1891, she entered the Knox family around 1907 as a practical nurse attending Hiram Knox, the son of Colonel William Hiram (Will) Knox, the Wisconsin patriarch of the lumbering family.

In the ensuing years at the Knox sawmill town in Polk County, Lillian married Hiram after his first wife died of blood poisoning, Hiram's teenage son was murdered, and Colonel Knox succumbed to a heart attack.

By 1917, Lillian was running a new Knox sawmill at East Mayfield, near Hemphill, while her husband spent most of his time hunting with his friends. The *American Lumberman* magazine called her one of the most influential women in the U.S. lumber industry and President Woodrow Wilson praised her for her work in producing ship timbers and selling war bonds during World War I. During the war, the Knox family had a worth of millions in industrial property, a railroad, oil and gas properties, homes in several cities, tens of thousands of acres of timberlands and other investments.

The death of Hiram Knox in 1922 left his wife a pariah in Texas social circles, and the Dallas bludgeoning death of her mother-in-law, Mary Knox, in 1937 thrust her into a new murder investigation, a string of check forgeries and employment at a Dallas hotel known for its association with criminals.

Eventually released for lack of evidence in the Dallas murder case, Lillian turned again to forgeries and spent four years in a federal reformatory for women. Released in 1944, she began using a series of aliases to develop confidence schemes in Illinois.

In 1966, at the age of 75, she was placed in an Illinois mental hospital, where she died and was buried in a pauper's grave, penniless, separated from her children, and a thousand miles from the Texas forests which made her a millionaire in her twenties.

In Texas, where Lillian Knox's story was told largely in oral histories for eighty years, she became the subject of legend and folklore—a strange mixture of an angel of mercy and an accused murderess. To newspapers, she became a magnet for scandal. To historians, she became an enigma.

The complete life of Lillian Knox had been largely overlooked by written history until 2002 when Texas historians Bob and Doris Bowman crossed paths with Illinois attorney Edward J. Barrett. As a boy, Bob Bowman lived with his family at East Mayfield, and had included a chapter on Knox Lumber Company in his 1988 anthology of East Texas ghost towns. When he was invited by the East Texas Historical Association to deliver an historical paper on the 1922 murder of Hiram Knox, the Bowmans began to dig deeper into the Knox legend.

Their work brought them in touch with Barrett, who had been collecting and recording his experiences and conversations with Lillian Knox since 1957 when he met her in Chicago's Cook County Jail. Barrett not only became Mrs. Knox's attorney, but a close confidant. As he approached retirement age, Barrett began working on a book manuscript of Knox's life.

When the Bowmans and Barrett compared their work in 2002, they found they shared a common fascination with Lillian Knox and soon became collaborators to produce the only book ever written about this mysterious woman.

The life of Lillian Knox is so improbable that it could have been written by a novelist. But the actual facts of her life—recorded in hundreds of documents in Texas, Louisiana, Illinois and Wisconsin—are enough to make Lady Bountiful one of the most controversial and remarkable women in Texas history.

Table of Contents

Lillian Knox in a formal portrait at the peak of her fame between 1917 and 1919. *Photo courtesy, Texas Forestry Museum.*

1

A STANDING OVATION. The evening of Saturday, June 10, 1922, was the finest night of Lillian Knox's life.

The thousands of people who packed Temple Auditorium in rural Sabine County had arrived for what the *Houston Post* described as "one of the greatest social events in East Texas."

The auditorium was decorated to create "a panorama of the most beautifully blended colors" and were "elaborate in every respect," reported the *Post*. A popular band had been imported from Beaumont, Texas, to provide music for a dance. The refreshments served "were most delicious and of a wonderful variety," added the *Post.*

The guest of honor was dressed resplendently in a new gown purchased for the occasion. Seated on the auditorium stage— surrounded by state officials, community leaders and titans of the lumber industry—she beamed with a broad smile, obviously pleased with what she saw.

Her peers in the industry had traveled to Sabine County

from Houston, Dallas, Beaumont, Lufkin, Kansas City, Shreveport and other distant cities to pay their respects. Private railroad coaches owned by the dignitaries stood on the Knoxes' private rail spur near their home.

Knox Lumber Company, Sabine County's largest employer, was founded in 1913 by Colonel William Hiram (Will) Knox, a Wisconsin lumberman who moved to Texas around 1900. Located in the village of East Mayfield, on the western edge of Hemphill, the mill had a capacity of 200,000 board feet every ten hours and employed 400 to 500 men and women in 1918. When he died in 1913 while the mill was being built, Colonel Knox's son and daughter-in-law, Hiram and Lillian Knox, steered the plant to completion and began producing lumber and shipbuilding timbers.

In its June 8, 1918, issue, the *American Lumberman* described Mrs. Knox "as the most remarkable woman executive in the lumber industry." She and her husband Hiram comprised "a real team," said the *Lumberman.* "Here are a husband and wife who divide the responsibilities of running a several million dollar sawmill property so that neither trespasses on the other. Mr. Knox is president and superintendent of operations, Mrs. Knox secretary and general manager."

But the *Lumberman* focused its article largely on Lillian: "At the handsome office of the Knox Lumber Company, Mrs. Knox is very much the dominant figure. It is she to whom every man in the mill, from bookkeeper to negro laborer, comes for advice or help...her office hours are as regular as those of her bookkeeper and she often goes back at night."

The *Lumberman* said Mrs. Knox "possesses more than executive ability. She is a rare organizer and a master of details."

Her skills were not restricted to the sawmill; she had raised three young boys and, according to the *Lumberman*, "her home is so well organized from housekeeper to second maid that it gives her no worries."

When the *Lumberman* article appeared, Mrs. Knox was also in charge of 152 acres of war gardens for World War I, an answer to President Woodrow Wilson's appeal to raise more food in 1918 than in previous years. To supplement the gardens, Mrs. Knox built a canning factory. "Employees (of Knox Lumber Company) bring the products of war gardens and get them canned at no cost but the wholesale cost of the cans. And all next winter, these employees will be eating vegetables raised by their own hands instead of paying fancy prices for imported foods," said the *Lumberman*.

To further support the war effort, every employee at the Knox sawmill was a subscriber to Liberty Bonds and Thrift Stamps. Said the *Lumberman*: "The Knox community...is one of the most patriotic in all Texas. There are over 1,200 members of the Red Cross, from Knox company families, and one of Mrs. Knox's young female protégés completed the knitting of fifteen sweaters for the soldiers from January to the first of April."

Mrs. Knox served as Sabine County's food administrator and when the Second Food District of Texas decided to observe March 1, 1918, as "Food Observation Holiday," she decided that she and her family would celebrate the holiday by closing down the sawmill and staging a parade through downtown Hemphill. "Mrs. Knox and her husband were in the lead, mounted on spirited horses, and the employees of the mill and their families followed in floats, automobiles and wagons." The parade—which flowed around the Hemphill courthouse square—consisted of 14 automobiles, 20 wagons and carts loaded with logs, and six decorated floats bearing "pretty girls and patriotic banners," said the *Lumberman*. Rows of employees bearing banners brought up the rear.

During the influenza epidemic of 1918, when Americans died by the thousands, Mrs. Knox employed physicians and nurses to care for the sick in Sabine County. She was also among Texas' early leaders of women's suffrage, donating $200,000 to the cause. She chaired the Fourteenth Senatorial District of Women's Suffrage, as well as the East Texas chapter of the Texas League of Women Voters.

For her war efforts, and the sawmill's ability to produce heavy timbers for the construction of warships in the Texas Gulf Coast's shipyards, Mrs. Knox was sent personal congratulations from President Wilson. "In some items of big ship timbers," said the *Lumberman*, "the Knox mill is declared to have surpassed all other mills in this section of the South."

The Houston Post said nearly 3,000 people were present for Mrs. Knox's tribute in the Temple auditorium, the largest assembly of people to attend a social event in Sabine County history.

County Judge J.B. Lewis, who had been designated by the event's planners to make the principal address, rose from his seat, took his place at the dais and smiled warmly at 31-year-old Lillian Knox, a diminutive, attractive person with a strong, but feminine face. Slightly on the plump side, she had piercing, determined eyes, a pug nose, black hair worn in a short, off the shoulder style, and a firm mouth. She wore jewelry around her neck, on her wrists and fingers, and on the bodice of her dress.

"This fine and gracious lady," Judge Lewis began, "has been the greatest individual benefactor Sabine County and its people have ever had." He recalled her numerous works of charity to the poor and said she had "tried to keep the public from knowing of her bountiful donations because of her unselfishness and her desire to not have those she had helped humiliated by having the world know they accepted her assistance."

On the verge of tears, he told how Lillian Knox had sent "many poor boys and girls through school and scarcely any one knew it but them." He recalled that she had used her skills as a nurse during the influenza epidemic of 1918 and, when she was inundated by patients, she obtained "all the available trained nurses she could from nearby towns" and placed them in the homes of her employees at the East Mayfield sawmill.

Further, he said, Lillian organized soup kitchens for the benefit of the ailing and spent thousands of dollars to staff them. "Not one dollar was charged to her employees," continued Judge Lewis.

He recounted instance after instance of orphan homes which Mrs. Knox had assisted "by furnishing lumber to build dormitories for the little children" and made significant cash contributions as well. He spoke of many other acts of kindness and especially "things done for the benefit of Sabine County and its citizens."

During Judge Lewis' speech, said the *Post,* "there were many eyes of Sabine County's best men, women and children seen to become wet with tears." His speech was interrupted frequently by applause.

Lillian Knox, he concluded, is truly "our Lady Bountiful." The audience stood and responded with prolonged applause.

When the applause had subsided, the evening's guest of honor rose from her chair and advanced to the center of the broad stage. She was greeted with "long and continuous applause," said the *Post*, "and finally the entire audience rose to its feet, bowed to Lillian Knox's presence, and seated itself."

Tears welling in her eyes, Lillian delivered "a brief but touching reply," reported the *Post*. The audience once more broke into a storm of cheering and applause "in recognition of their great respect, love and and admiration for a most noble and kind-hearted women of rare and useful attainments."

It was the greatest night of Lillian Knox's career and almost every influential person in Sabine County, state and federal officials, and leaders of the lumber industry in Texas and Louisiana willingly shared it with her.

The one notable exception was her husband, Hiram Knox, who remained at home.

2

THE KNOX EMPIRE. While Lillian Knox was outgoing, generous and lively, her husband Hiram was the opposite—retiring, conservative and cautious in making decisions. But there was a good reason for Hiram's reticence. He was the product of a demanding and aggressive father whose will was seldom challenged, especially in his own family.

Colonel William Hiram (Will) Knox, a timberman and sawmill owner in the Lake States, had been an entrepreneur and adventurer since he was a young man. All of his life he had followed the American frontier as it presented opportunities for land speculators, lumber producers and railroad builders.

Knox's brother, Samuel Goodhue Knox, came from Pembroke, New Hampshire, to Wisconsin at the age of 23 in 1855, opening a country store in the village of Oxford in Marquette County. William, 21, arrived in 1856 to join his brother's store, at which time they added lumber products to the business. Descendants of Scottish ancestors, their maternal grandfather had fought in the American Revolutionary War. William Hiram Knox was born November 28, 1834, two years after Samuel's birth on December 11, 1832.

In 1862, the brothers sold their Oxford store and built a grist mill. Samuel married Eliza Babcock that year. William also met Mary S. Axford at Oxford and married her on July 14, 1865. Mary, who had been born in New Jersey on January 6, 1846, came from a prominent Great Lakes family; her grandfather was John S. Axford, a member of the Michigan Constitutional Convention. The town of Oxford was named for the Axford family, but careless neighbors, through mispronunciations, gradually changed the name to Oxford. Members of the Axford family believe William Knox's early lumbering investments were funded by money inherited by Mary from her family.

William and Mary had two children, both apparently born in Wisconsin. Hiram was born on August 18, 1866, and Grace Erma Knox arrived in the family on June 17, 1870.

In 1862, following the outbreak of the Civil War, the Knox brothers moved to Montana to avoid the draft and explore new frontier wealth. They remained in the West until the war's end, returning in 1867 to Wisconsin, where they purchased timberlands and built a sawmill at Crooked Ripp, Wisconsin. Cutting their timber in six years, they closed the mill and moved to Stevens Point in 1873, building another sawmill, which they sold in 1877 shortly before a fire destroyed the facility.

Moving to Wausau, they constructed still another sawmill in 1878 and began producing lumber as Knox Brothers Company, cutting timber from their large holdings in Lincoln and Oneida counties and shipping it down the Wisconsin River to their Wausau mill. William Knox was also the president of Wausau

Lumber Company and had two partners, James McCrossen and D.L. Plumer.

Old Wausau city directories indicate that William and Samuel were in business in 1886 as Knox Brothers, a wholesale business dealing in lumber, shingles and pine lands. Their offices were on Main Street.

On January 1, 1881, joined by McCrossen, the Knox brothers purchased 2,960 acres of land from Francis Palms in Price County, Wisconsin, for $24,000. In November of the same year they contracted with the Wisconsin Central Railroad to cut the pine timber on 2,400 acres of forestland owned by the railroad. The Knox brothers, however, were no strangers to the region. They had been cutting timber in the Willow area, about ten miles east and north of their new landholdings, for many years.

The lands the Knox brothers purchased were covered with a variety of species, including hemlock, white and yellow birch, tamrack, spruce, balsam, rock elm, butternut, red oak, hard and soft maple, basswood, white and black ash, cedar and aspen. But the vast forests of white pine were more coveted by lumbermen.

In *"The Culture and Community of Knox Mills, Wisconsin,"* Joyce T. Bant described the land: "Through this unbroken forest, in southeastern Price County, a small stream, Long Meadow Creek, now known as Knox Creek, flowed into a branch of the Manito, or Spirit River. This forest land had been virtually untouched by humans. Native Americans followed trails north and south of the area, but not through this area. An occasional hunter followed Long Meadow Creek, as is evidenced by a flintstone arrow found along its banks a long time ago by a Knox Mills farmer, but other than that, there is no other evidence that anyone ever set foot on this land until July 1, 1864. On that day, the solitude was broken when a surveyor, with the help of two chainmen and an axman, began surveying Township 35 North, Range 3 East. The task was completed that year and the land was opened for settlement. Some of the land was granted to the railroad; some was reserved for homesteads; and the rest was offered for sale."

Immediately after the land was put on the market, speculators arrived to grab the land with intentions to hold it until settlers arrived when they could sell it at a profit. During 1867 and 1868 Francis Palms had obtained the lands bought by the Knox brothers, along with other holdings in Price County.

The lands Palms obtained in Price County were granted to him by acts of Congress between 1847 and 1855 for military service. Service with the Union Army in the Civil War did not necessarily entitle Palms to so much land, so he arranged for the land to be granted to other soldiers, some of whom were dead, and then had it assigned to him. Palms also acquired lands through an 1862 Congressional act which granted land for agricultural purposes. His acquisition of lands in Wisconsin and Michigan gave him the distinction of being the largest landowner in the Midwest, perhaps the entire United States, at the time.

After the Wisconsin Central Railroad was built into Price County in 1873, the railroad and the State Board of Immigration sent agents to Sweden—which had a climate similar to Wisconsin—to persuade people to come to Wisconsin and settle on the lands. The appeal focused on poor people and claimed that all the settlers needed were a couple of dollars and an ax to build a home, provide fuel and clear land for a garden.

The railroad land could be purchased at a price ranging

from two to six dollars an acre with one-quarter down and seven percent interest over a term of several years. A good log house cost about forty dollars, including labor, and took four to five days to build. During the logging season, an unskilled settler could work for a logging company for twenty dollars a month. A man with a good team of horses or oxen could make sixty-five to seventy-five dollars a month and board.

Like other transient lumbermen working on the frontier, the Knox brothers had a single motive in acquiring the lands. They wanted to harvest the white pine, produce lumber and move to another territory.

And, like other lumbermen, they often fought with local authorities over the amount of taxes they should pay. In at least one case, the Knoxes—joined by McCrossen and the Wisconsin Central Railroad—carried on a three-year dispute over taxes in Price County. The Knox brothers claimed the Swedish settlers controlled the tax assessors, who were elected because they intentionally omitted the value of improvements made on the settlers' lands and improved farms. They also accused the public officials of discrimination against non-resident taxpayers such as the Knoxes and other logging companies.

Will Knox later helped elect several assessors who would be friendly to him. Assessor Patrick Murphy had been a foreman in Knox's logging camp and assessor W.T. Goodhue was Knox's cousin. Goodhue had been raised by Samuel Knox from the age of 13 after Samuel lost two small sons in 1873.

In addition to their lands in Price County, the Knox brothers owned holdings in Lincoln and Oneida County, which made them one of the largest logging operations in the area. During the logging season of 1878 and 1879—using a crew of only twenty-four men, three pair of horses, and a yoke of oxen—the Knoxes logged a million feet of timber in twenty-seven days. Before the season ended, they banked another two million feet of logs for shipment down river to their mill at Wausau.

Joyce Bant said the Knox brothers had reputations as good employers. "They paid their employees well; their wages...were as good or better than the scale that the Lumber and Log Owners Association of Ashland adopted for the 1888-89 season," she said.

Even though they were transients who owned a sawmill outside the area, the Knoxes were involved in the community. They lived and worked in Price County and socialized in the Spirit River area, south of their timberlands. Knox also operated a large farm to feed his logging crews. However, the Knox brothers apparently continued to make their home at Wausau. City directories listed all of the Knox families—Samuel and his family and William and his family—living on Franklin Street between 1883 and 1890.

In the summer of 1890, the Knox brothers' big sawmill at Wausau burned. Rather than rebuild the mill, William and Samuel agreed to divide their interests and move in different business directions. On August 1, 1890, Samuel sold his personal and real property, as well as his unexecuted timber licenses with the Wisconsin Central Railroad, to William for $81,500 and moved to Prentice, Wisconsin, where he built a mill.

William chose to remain in the forestlands he owned in Price County. To do so, he moved the usable machinery from the Wausau mill to a site on Long Meadow Creek. There, he built a mill pond to hold logs until they could be converted into lumber.

Knox persuaded the Minnesota, Sault Ste. Marie and Atlantic Railroad to build a branch line from the town of

Brantwood to his new millsite on Long Meadow Creek. When the line arrived, Knox completed construction of his mill in June of 1891. A town known as Knox Mills also grew up around the sawmill and on August 22, 1891, a post office was established in the community with Hiram Knox, William's son, as the first postmaster.

The town was split by Long Meadow Creek. On the east side were two boarding houses, six houses with two families living in each, a lumber company office, a company store and the post office. On the west side were twelve houses for the employees.

Joyce Bant said the east side of the town was known as Frog Town because of the large number of frogs present around the mill pond. The west side was called Pig Town because most of the occupants of the company houses raised pigs for butchering.

According to Bant, the Knox mill was the last of the pine lumber mills to be built in Price County, following other mills constructed in 1874, 1882, 1884, 1888, 1889 and 1890. While the Knox mill had an annual output of 150 million board feet between 1891 and 1895, it was not the largest. That distinction went to the J.R. Davis mill at Phillips, which had an output of 570 million board feet between 1882 and 1892.

"During the time that Knox ran his mill at Knox Mills, he always paid his men with currency. Newspapers quoted him as saying he abhorred the script coupon system of payment for labor (the custom of the day in lumbering towns). Instead of having cash to pay for the things they needed, coupons were issued the men to be used at the company store in exchange for their purchases. Although most of the things that were needed were provided by the company store, it (Knox's decision to pay in currency) did give the men more freedom to purchase things that the company store did not provide," said Bant.

As with most lumbermen of the era, Knox was interested in politics. At Knox Mills, he ran for chairman of the Board of Supervisors in the town of Brannon and served from 1891 until 1896. During that time, his entire ticket ran the town. He also took an interest in national politics. Although he was a staunch Democrat, he joined the "bolters" in the election of 1896 when William Jennings Bryan was the Democratic candidate. Knox publicly opposed Bryan's proposals for free, unlimited coinage of silver and taxation. One of the Democratic proposals was an income tax of two percent on incomes over $4,000. "Although known as a square and honorable businessman, whenever finances were an issue, Knox was there to defend his interests," said Bant.

On September 3, 1891, young Hiram Knox married Grace Eva Briggs in Wausau. The daughter of Carlton Wilberton Briggs, a lawyer and a former Civil War captain, and Eva Grace Fraser, a lineal descendant of Sir Walter Raleigh, Grace Eva was born March 18, 1872, at Oxford, one of five children in the Briggs household. She was educated in the Wisconsin Rapids public schools and graduated from Howe High School in 1888. She taught in a Wood County rural school for three months that fall before accepting a position as head accountant with a manufacturing plant at Centralia (now Greater Wisconsin Rapids), where she worked two years. She then worked for Spafford and Cole, a merchandiser, at Rhinelander, Wisconsin. A biographer said Miss Briggs, "in addition to her duties as accountant, proved to be an able assistant in selecting millinery and dress goods of a higher grade than that formerly stocked by the company." Eventually, she was employed at Wausau, where she met Hiram. Their only child, William Hiram (Willie), was born on June 14, 1894, while Hiram and Grace Eva were living at Knox Mills.

By the summer of 1895, most of the white pine had been cut on Knox's lands and it was apparent Knox Mills would never amount to anything more than a mill town at the end of a railroad line. The elder Knox began advertising his cutover lands as being suitable for small farms with 40 to 80 acres. E.H. Hobe, an immigration agent for the Sault Ste. Marie and Atlantic Railroad, was employed by Knox to encourage people to settle on the lands.

On June 23, 1896, the planing mill at the Knox sawmill was destroyed by fire, a frequent occurrence among sawmills. Knox had no insurance and sustained a loss of $6,000. By the end of the year, Knox had sold his mill to a party in Superior, Wisconsin, and his lands to Hobe. All of the Knox mill's machinery was torn down and shipped to Superior, where Knox and his son supervised the rebuilding of the mill.

During the transition, Knox indicated to his family that he would like to spend his summers in Superior and built a summer home there, but he later announced his intention to travel with his wife Mary to a warmer climate for winter.

The move may have been precipitated by the decision to place their daughter, Grace Erma Knox, (not to be confused with Grace Eva Briggs Knox, young Hiram's wife) in a New Mexico tuberculosis center, where she died on September 24, 1900.

While scouting for a winter home in Texas, Knox began to hear about the vast, virgin yellow pine forests growing between the Trinity and Sabine Rivers in the eastern portion of Texas.

Around the turn of the century, there was considerable rivalry between white pine lumber produced in the northern

U.S. and the expanding yellow pine industry in the southern U.S. In 1900, the *Southern Lumberman*, a trade publication, alluded to the rivalry: "The history of the Southern yellow pine and the northern white pine trade of twenty years past, if it could be briefly and intelligently written, would be highly interesting as showing what a strong fight the manufacturers of each have made for expansion of trade territory of each wood. The encroachment of the yellow pine upon the territory of the white pine dealers once considered their own, has been steady..."

As the new century approached, Knox and his son, who had already made a fortune in white pine, moved into the backyard of the competing yellow pine producers. Accompanied by their families, they selected Dallas as their base of operations.

Even though he was already in his sixties, the elder Knox wasn't quite prepared for retirement. He was ready for a new challenge in Texas.

A winter scene in Wausau, Wisconsin, with the William Hiram (Will) Knox home looming at the end of the street. *Photo courtesy, Wausau Historical Society.*

The Wausau, Wisconsin, home of William Hiram (Will) Knox, as it appeared in the l890's. The children were apparently unrelated to the Knox family. *Photo courtesy, Wausau Historical Society.*

William Hiram (Will) Knox stands before a stack of lumber shortly after he arrived in Texas. *Photo courtesy, Polk County Memorial Museum.*

3

A NEW EMPIRE. Around the turn of the century, Colonel William H. (Will) Knox moved from Wisconsin, settling in Dallas with his wife Mary. Accompanying them were their son Hiram, his wife Grace Eva, and their only child, William (Willie) Hiram Knox, who was enrolled in a private school in Dallas and later attended Peacock Military Academy in San Antonio.

The colonel had good intentions of retiring. "But he got tired of doing nothing and went back into the lumber business in East Texas," said a relative. While in Dallas, the aging lumber baron became enamored of reports of millions of acres of cheap timberland southeast of Dallas.

The forests had greeted the first settlers coming into East Texas as early as 1812, but a lack of dependable transportation had discouraged lumbermen from investing in large sawmills in the region's interior, although some mills were built at Orange on the Texas Gulf Coast and utilized logs floated down the Sabine River from Newton and Sabine counties.

The catalyst to the development of the forest products industry was the construction of the Houston East and West Texas Railroad (locally known as "Hell Either Way Taken ") from Houston to Shreveport, a distance of nearly 225 miles, between 1876 and 1886. As it moved in a northeasterly direction through the midsection of East Texas, the railroad left in its wake the creation of dozens of new sawmills, many of them built by transient or "gypsy" lumbermen anxious to capitalize on trees so cheap that farmers were willing to give them away just to have their lands cleared for crops.

East Texas soon became a spider web of logging trails, tram railroad lines, sawmill towns and logging camps. Most of them were connected in some fashion with the Houston East and West Texas Railroad or connecting railroads lines, which gave the lumbermen a way to move their lumber to distant markets.

Flush with money realized from the sale of his Wisconsin timberlands and the sawmill at Knox Mills, Wisconsin, Knox quickly saw the potential of the lumber business in Texas. As he traveled throughout East Texas, he was informally bestowed with the honorary title of Colonel—the Old South vernacular for a respected individual—but he soon acquired a reputation as a stern, cantankerous Yankee capitalist with an incendiary temper.

Colonel Knox apparently spent considerable time in Texarkana, then the crossroads of the lumber industry for Texas, Louisiana and Arkansas, largely because the Missouri Pacific, Cotton Belt and other major railroad lines crossed in the city, allowing lumber to be moved by rail in almost any direction. From Texarkana, Colonel Knox and his son Hiram operated a lumber business and established a large retail lumberyard at Mission,

Texas, in the Rio Grande Valley, described as the first of its kind in Texas. They also began exploring the East Texas forests, including those growing in Polk County, about 50 miles north of Houston.

In April of 1901, Colonel Knox and Hiram purchased a block of about 10,000 acres of timberland in the vicinity of what is now Soda, a small community east of Livingston in Polk County. He also purchased a home in Hidalgo County for his family.

The Soda community was originally known as Bluff Creek for a small creek running nearby. Established about the time of the Civil War, Bluff Creek had a post office between 1860 and 1866 and was a center for agriculture and small lumbering operations. It also had a small school. In 1898, the town acquired another post office, called Soda, a name derived from the first letter of four names submitted to postal authorities by residents.

Named for President James K. Polk, Polk County was one of twenty-three new counties founded in 1846 by the Texas Legislature when Texas was annexed to the United States. The county's boundaries closely followed those of the old northern division of the Liberty Municipality, a subdivision of the Department of Nacogdoches established by the Mexican government in 1830 before Texas won its independence and became the Republic of Texas. A part of Polk County west of the Trinity River became San Jacinto County in 1870. The present boundaries of Polk County were fixed in 1875 when a portion of Trinity County was annexed. Livingston, formerly Springfield, became the county seat in 1846. The development of the county was spurred by transportation improvements, including shipping on the Trinity River and the completion of the Houston,

East and West Texas Railroad.

In November of 1904, Lucian Dewalt, a Civil War veteran who owned land in the Soda area, wrote his brother that "we're going to have a railroad soon...and a big sawmill put up by the Knoxes...so I'll have a place to sell my timber." In May of 1904, Dewalt wrote that "they are putting down rails" and "will soon put up the sawmill."

In 1905, Colonel Knox and his son Hiram completed a large steam-powered sawmill near Menard Creek, where ample water was available. The mill initially employed about forty men and spurred a temporary rebirth of Soda, which expanded to include a general store, a hotel and other new buildings. The lumber company was created as Knox Lumber Company and the settlement around the mill was known variously as Knoxville and Knox. The mill, which had a band sawmill, a planer and dry kilns, had a capacity of 80,000 board feet of lumber in 1905, but was producing 150,000 board feet a day by 1906.

The Beaumont Enterprise described the Polk County sawmill in 1905: "The big sawmill at Knoxville is about ready to commence making lumber. Not Knoxville, Tennessee, but Knoxville, Polk County, Texas, seven miles east of Livingston. Mr. Knox is the owner of the mill and also the tram or railroad that runs to Knoxville, which is in the pine woods at the terminus of the railroad. He owns several thousand acres of fine pine land, all paid for, and if he doesn't get rich, it's his own fault..." The newspaper said the mill would eventually provide manuacturing, logging and related jobs for 250 men.

Frank Angell, who had moved from Wisconsin to Texas to rejoin the Knoxes in the sawmill venture, was the mill foreman. Other key employees included saw filer M. Barnhart, sawyers

Fred Bonds and a Mr. Fryer, loggers Whit Dillon, Walter Murphy, Charlie and Ed Duff, logging foremen Oscar Rhoden and John Winger, block setters Fred Newton and Bea Dillon, yard foreman and log breaker Walter Williams, lot feeders and overseers Henry Dillon, Lon Oliver and John Dison, carpenters Clifton Packer and Bill Chambless, bookkeeper Elbert C. Wilson, millwright Joe Zimmerman, blacksmith Clifton Packer, nightwatchman Whit Dillon, shipping clerk Ed Malone, log scalers John Dillon and John Malone, locomotive engineer L.H. Placker, and railroad conductors John Frisby and Pete Bailey. Dr. Soloman Adams served as the company doctor and Samuel B. Kelly was manager of the company store and post office during the mill's early years.

On December 18, 1903, Colonel Knox chartered a common carrier shortline railroad, the Livingston and Southeastern, which soon connected Soda and Knoxville with the mainline railroad, the Houston, East and West Texas Railroad, at Livingston. Owned by Knox Lumber Company, the railroad's directors were Colonel Knox, his son Hiram, C.W. Hawley and W.F. Robertson, all of whom were listed as residents of Dallas County; C.A. Noblett and L.R. Gerlach, both of Polk County; and James E. Miles of Angelina County.

The first seven miles of the railroad were built in 1904 to connect the Knox sawmill with the HE&WT. In 1908, another three miles was added, completing the company's ten miles of track and connecting the L&SE with the Beaumont and Great Northern line, giving the L&SE connections with two mainline railroads at Livingston. The railroad not only carried lumber and other freight between Knoxville and Livingston, but ran a one-coach passenger train into Livingston each day.

In 1907, Colonel Knox, who was involved in local and

state politics, offered a free train ride to all men of the community if they would join him in Livingston to hear a speech by Joe Bailey, a candidate for the Texas Senate. A photograph of the crowd of men was taken near the HE&WT depot and the Oleander Hotel in Livingston.

Hiram Knox, accompanied by his wife, Grace Eva, and young son Willie, moved to Knoxville so he could operate the lumber company for his father. They moved into a large four-room company house described as "a bungalow" by Lillie May Creel, a former Knoxville resident. The house had four rooms and two large hallways and stood with other dwellings on a square of buildings which also included the mill, a store, railroad depot, other dwellings, a school and a Baptist church. Almost everyone in the community belonged to the church and oldtimers said attendance was almost 100 percent. The school provided nine grades with three teachers at its peak.

Colonel Knox, who continued to maintain homes in Dallas and Mission, often came to Polk County, where he earned a reputation for intemperate action. People were prone to distance themselves from him and his business operations because of his temper.

Colonel Knox reportedly shot and killed a white man in June of 1906 on a Livingston street because the man had abused one of his black employees at the sawmill. The colonel was indicted on a murder charge June 12, 1906, and released on a $1,000 bond. He was tried in Livingston on December 7 of that year. After the state had put a few witnesses on the stand, District Attorney J.L. Manry asked the jury to return a verdict of not guilty on grounds of self defense. The *Polk Country Enterprise* reported: "Mr. Knox is a prominent mill man of the county and his friends are rejoicing on his exoneration."

Colonel Knox's temper was in keeping with his appearance. A photo of the old timber baron, taken while standing in the middle of a railroad track around 1911, captured him with a scowling countenance—a near-perfect match to the face of the black bulldog at his feet.

Colonel Knox was also quick to defend what he felt were his interests and property, occasionally resorting to lawsuits in addition to physical action. In 1902, a year after he purchased his Polk County lands, the colonel filed trespass charges against S. Collins and Lima Lumber Company in a Polk County court, alleging that Collins had cut timber worth $15,000 on a tract of Knox land, and was holding the timber at the Lima sawmill in Polk County. The lawsuit was settled on June 18, 1903, with Knox receiving the value of the timber, but it amounted to less than $2,400.

The Knoxville community seemed to have more than its share of tragedies. In July of 1909, another shooting left a traveling peddler named Fields seriously wounded from shotgun wounds at Knoxville. W.H. Cubley, who did the shooting, surrendered himself to peace officers in Livingston. Fields was sent to Humble on the afternoon train. The same month, a L&SE railroad engine struck a steel rail lying across the tracks near the Menard Creek bridge, resulting in a derailment of a log car carrying several men. Two men were injured and rushed to a hospital in Houston for treatment.

In 1909, a staff writer for the *Polk County Enterprise* toured the Knox mill and reported: "We found Mr. W.H. Knox, head of the Knox Lumber Company, who showed us many courtesies while at his saw mill town. The saw mill plant is an excellent one, being modern in construction and convenience. The mill was stopped for some additional improvements, but was about ready to start again. Mr. Marshall is mill foreman and Mr. Jamerson is saw filer. The planer was running with full force. Its foreman, Mr. W.W. Owens, was on hand, keeping everything in splendid shape for turning out fine grade lumber. Mr. Eli Malone, the shipping clerk, was at his post of duty and everything at this saw mill plant was moving along like clockwork. Mr. Frank Angell, next in charge under the company, and Mr. E.C. Wilson, general bookkeeper, were at their places of duty. Mr. Hiram Knox, junior member of the company, was in the Kingsville country in the interest of the firm."

On May 13, 1909, the *Enterprise* reported that Hiram Knox used the Livingston and Southeastern Railroad to carry a gathering of people from the HE&WT connection at Livingston on "a moonlight picnic" to the sawmill. The newspaper reported that the guests played games at the company hotel "until supper was announced...lunches had been prepared by the ladies in the crowd and it was indeed a feast that was enjoyed by the entire party." Lillie Mae Creel remembered the community as "a nice little town."

Around 1907, while traveling in Houston, Hiram Knox, became ill with pneumonia and his family secured the services of Lillian Marshall, a divorced but childless practical nurse who lived in Houston, to sit with Knox—a decision that would influence and haunt the Knox families for nearly a century. Lillian apparently helped Hiram recover. An article in the *Polk County Enterprise* on August 1, 1907, reported that "Mr. Knox, who has been sick for the past few weeks, is able to be up again."

Born Lillian Marguerite Marshall in the Comanche

County settlement of DeLeon on November 15, 1891, Lillian was one of four daughters and a son born to George and Daisy Marshall. The father, a seed and livestock feed salesman, spent most of his time on the road, leaving his wife to raise Lillian, Clara, Fannie, Evelyn and Alex.

Daisy Marshall was a God-fearing, Bible-quoting, no-nonsense disciplinarian, especially with her daughters. She often lectured her daughters on sin, damnation and proper conduct, especially around men. Clara was the oldest, Lillian the youngest, and Fannie and Evelyn in between. Alex, the only son, was indulged by his parents.

School for the Marshall children was a renovated barn with one teacher for all grades. Most of the school's children advanced only to the eighth grade because there were responsibilities and chores at home. Comanche County was poor, sparsely populated and far removed from towns with any appreciable population.

But when Lillian was ten years old, the Marshalls' hard-scrabble life in DeLeon ended. At Daisy's insistence, George agreed to quit his sales job and move the family to Texarkana, an up-and-coming town on the Texas-Arkansas border. On the east was Miller County, Arkansas; on the west was Bowie County, Texas.

In Texarkana, Daisy could care for her widowed father, whose eyesight had begun to fail. Loading its possessions in a horse-drawn wagon, the family made the 300-mile journey to Texarkana and moved in with Daisy's financially comfortable father. George soon found another job in Texarkana and the family settled down in a new, larger and more pleasant community. Oldtimers in Texarkana remembered Lillian Knox as "one of two little girls who accompanied a blind man known as Blind Nelson" around Texarkana, according to the *Texarkana Gazette* in a 1937 story.

While her sisters entered into successful marriages, Lillian was less fortunate. At the age of 15, she quit high school and married Roy J. LaRue, 24, of Texarkana on July 23, 1906. The childless marriage lasted less than two years. LaRue was remarried in 1910, this time to Effie M. Fenley, 35, his step-grandmother.

Sometime between 1906 and 1907, George and Daisy Marshall moved the family to Houston, where they settled in the city's old Sixth Ward and Lillian became a practical nurse, according to the *Houston Chronicle* on January 23, 1923.

Entranced by her youthful beauty and her common sense, and elated that she had nursed his only son back to health, Colonel Knox persuaded Lillian to leave her nursing profession and accept a position in his lumber business, where she was installed as his personal secretary, despite her lack of secretarial training, and later as his accountant.

With a decent salary and Colonel Knox's confidence, Lillian immersed herself in the family's lumbering empire. She was soon handling the Colonel's business and personal affairs and began learning the lumber business from the ground up. As the Knoxville mill grew, Lillian assumed more and more responsibilities, frequently exceeding those of Hiram. She often accompanied the Colonel on his trips to Mission and Houston.

On June 10, 1908, Hiram Knox's wife, Grace Eva, died in their home at Knoxville, leaving him to raise the only child, 14-year-old Willie. Grace Eva was buried in Forest Hill Cemetery in Livingston. Lillian had also nursed Grace Eva, but her skills did not work this time.

On November 16, 1910, Lillian and Hiram were married

with the Colonel's blessings in a church ceremony, followed by a lavish reception in Texarkana. Lillian was only eighteen, according to her reported birth date of November 15, 1891, and Hiram was forty-two. Despite the differences in their ages, Colonel Knox was delighted to have the beautiful and talented Lillian as his daughter. He also hoped she would help install some backbone into his unambitious son, who spent most of his time on hunting expeditions in South Texas.

Lillian's marriage came only nine months after suffering the loss of her mother, Daisy Nelson Marshall, on March 16, 1910. Daisy was buried in Woodlawn Cemetery in Texarkana beside two infants who died earlier. George Edwards Marshall died on July 12, 1901, at the age of two and Iola Gertrude Marshall passed away on September 29, 1905, at the age of three.

In a photograph taken shortly after their marriage, Lillian looked younger than eighteen and had a girlish, carefree grin on her pretty face. Hiram looked much older, but he, too, appeared happy. In the 1910 U.S. Census, Hiram and his father were listed as living in adjoining houses at Knoxville, although the elder Knox maintained his official residence in Dallas.

Following a honeymoon in Europe, Lillian and Hiram returned to Knoxville in Polk County, where young Hiram resumed running the sawmill, but in 1911 another tragedy struck the Knox family. On Saturday night, March 4, young Willie, now seventeen, was stabbed to death in a Knoxville home, resulting in murder charges against an employee of the Knox railroad. Willie was buried beside his mother's remains in Forest Hill Cemetery on Sunday afternoon, March 5.

While living at Knoxville, Lillian gave birth to her first child in the Brazos Hotel in Houston. According to the Knox family Bible, Hiram William Ambrose was born on August 17, 1911, followed by another son, William Hiram, on June 16, 1912. He was also born in the Brazos Hotel. Years later, however, Hiram William Ambrose located his birth certificate, which indicated he was born in 1909—a year before Lillian had married widower Hiram.

Later in her life, Lillian claimed she had been married previously to Chester B. Collins, a Lufkin lawyer who had represented the Knox interests in legal transactions. Collins was widely known as a "ladies man" and young Lillian was probably captivated by his glib tongue, suave demeanor and expensive suits. Lillian said she and Collins kept the union secret from Colonel Knox and lived apart.

Meanwhile, after only eight years of operations, the Knox lumbering enterprise—like other gypsy sawmills which had moved to Texas around the turn of the century—had overcut its timberlands and reserves. Only stumps remained on the land and the Colonel—unlike other Texas timbermen who envisioned future forests with the help of natural regeneration and manual tree-planting practices—saw no value in the land if it did not have large trees which could be converted into wide planks of lumber.

Colonel Knox told Hiram and Lillian he had decided to withdraw from Polk County and move his sawmill to Sabine County, where he had purchased 25,000 acres of virgin pinelands in Sabine, San Augustine and Newton County in 1912. By 1913, the railroad from Knoxville to Livingston was abandoned and the sawmill was removed and relocated to Hemphill in Sabine County. The *Polk County Enterprise* reported on November 20, 1913: "The Livingston and Southeastern Railroad is no more. The last rail was taken up and loaded on a car to be shipped away Thursday...the Knox Company finished cutting its timber in this territory about two years ago. The mill was removed several weeks ago to some point in Sabine County. The railroad material and equipment has also been shipped there for use in constructing another short line..."

As workmen, including some employees of the Knox sawmill and railroad in Polk County, began to build Colonel Knox's new sawmill near Hemphill, he surprised his family and business associates with another disclosure. He was divorcing his wife of 46 years, Mary Axford Knox, leaving her with a substantial divorce settlement allowing her to live in Dallas, far removed from Colonel Knox and his temper. In addition to the divorce settlement, Mrs. Knox was financially comfortable on her own as a result of an inheritance from her own family.

On June 26, 1913, as work continued on the new sawmill and another railroad at Hemphill, Colonel Knox died unexpectedly in his room at the Brazos Hotel in Houston at the age of 78. He was also buried in Forest Hill Cemetery.

The death of his father was a sharp blow to Hiram, who had returned with Lillian to Dallas while the new mill was being built in Sabine County. Even though he had learned the lumber business from his father, he lacked the zeal possessed by the old colonel. For most of his life, Hiram had lived in the shadow of his father, unable to overcome the dominance of a man known for his power, influence and temper.

In his will dated May 14, 1913, the colonel left his estate to Hiram, Lillian and their two sons—the children of Hiram and Lillian since their marriage. The colonel left Mary with $75,000 in securities, and additional bequests were made to his sister, Mrs.

John K. Sanderson and her infant son, Walter, of Lincoln County, Wisconsin. Hiram, now residing in Dallas, was named the executor of the estate and E.J. Mantooth of Lufkin was listed as the attorney. According to an account in the *Houston Post,* Hiram's inheritance amounted to an estimated ten million dollars in sawmill machinery, timberland, property in the Rio Grande Valley, several homes, investments, and more.

For Lillian, the colonel's death posed an opportunity. She knew Hiram's weaknesses and, even more important, she knew more about the colonel's business affairs than his son.

As Hiram's wife, Lillian Marshall Knox, the poor little girl from DeLeon, now had her grasp on a lumbering empire worth millions of dollars. She was Mrs. Hiram Knox, the leader of one of Texas' wealthiest lumbering families and one of the most attractive women in East Texas.

At the age of only twenty-two, she was prepared to assume the role of a millionairess in Sabine County, a sparsely-populated rural county sitting on the easternmost edge of Texas, just across the Sabine River from Louisiana.

Many former residents of Knoxville, site of a sawmill owned by the Knox family, are buried in Bluff Creek Cemetery in Polk County. *Photo, Bob Bowman.*

A group of Livingston residents stand beside a stack of logs at the Knox sawmill while touring the mill in January, 1909. *Photo courtesy, Polk County Memorial Museum.*

Lillian and Hiram Knox, following their wedding in 1910. *Photo, Texas Forestry Museum.*

Willie Knox, Hiram and Grace Knox's only son, died from knife wounds in 1911 at Knoxville in Polk County, Texas. *Photo courtesy, Merton Knox family.*

A logging crew pauses during its work in an East Texas forest shortly after the turn of the century. *Photo courtesy, Cecil McDaniel.*

Lillian Knox around 1915, a few years after leaving Polk County. *Photo courtesy, Mrs. Frank Hathcock.*

Hiram and Lillian Knox's first two children, Hiram William Ambrose, left, and William Hiram, were born when their parents lived at Knoxville in Polk County, Texas. *Photo courtesy, Merton Knox family.*

The Livingston and Southeastern Railroad, owned by the Knox family in Polk County, once traveled along this raised bed in the Soda community. *Photo, Bob Bowman.*

Menard Creek provided the water for the steam-powered Knox sawmill in Polk County. *Photo, Bob Bowman.*

4

THREE KNOX GRAVES. In 1913, as Lillian and Hiram Knox left Polk County—the scene of the Knox family's first lumber manufacturing operation in East Texas—they not only put behind them the remnants of a sawmill and a railroad, but the remains of three Knoxes whose deaths would portend a trail of tragedy for twenty-two-year-old Lillian.

During the less than five years Lillian lived in Polk County while working for Knox Lumber Company, she was on the perihery of the unexpected death of Grace Eva Briggs Knox, Hiram's first wife; the murder of seventeen-year-old William Hiram (Willie) Knox, Hiram and Grace's only child; and the death of Colonel William Hiram (Will) Knox—Lillian's aging benefactor, the patriarch of the family and the creator of the Knox lumbering empire in East Texas.

The three Knoxes were buried in Forest Hill Cemetery in Livingston, seven miles west of Knox Lumber Company's sawmill and the easternmost terminus of the Livingston and Southeastern Railroad, both created by Colonel Knox and Hiram in 1903.

The first to die was Grace Eva Briggs Knox, who had married Hiram September 3, 1891 (the same year Lillian Knox was born in Texas) in Wausau, Wisconsin, and accompanied him to Texas with their only child, Willie, when Colonel Knox summoned Hiram to help build a new lumbering operation in the virgin yellow pine forests of East Texas. Living in a company-owned house at Knoxville, Grace fell ill from tonsillitis and gradually grew worse during the ensuing hours. She passed away at 2 a.m. on Wednesday, June 10, 1908. According to two sources— an interview with former Knoxville resident Lillie Mae Creel and *Master Detective* in a January 1938, article on the Knox family—Lillian Marshall was nursing Mrs. Knox when she died. The *Polk County Enterprise* reported on June 11, 1908: "(Mrs. Knox) had been affected with tonsillitis and for the past week grew worse. On Tuesday, she was seemingly better, but that night got worse, reaching her heart, when the end came." Grace's death certificate said she died of blood poisoning. She was buried in Forest Hill Cemetery on the afternoon of June 11.

A family biographer wrote a somewhat different story of Grace's death. "Mrs. Knox died on June 10, at her home...on the day on which they had planned to leave for the East. She was about the house with friends as late as eleven in the evening, and passed away at three o'clock in the morning. Her death was caused by blood poisoning."

Grace was described by the family biographer as "someone who kept in touch with the world and its progress. Her personality made an impression not easily effaced. Her love and sympathy for people of all kinds and classes could not be surpassed. She was appreciative of the fine things of life."

Grace was an amiable, affable and cultured woman, reported the *Polk County Enterprise.* She also was an avid poultry raiser. In July of 1909, Hiram attended a poultry show at Houston to exhibit poultry raised by his late wife. The *Enterprise* reported on July 1, 1909: "Mrs. Knox, while living, took considerable interest and pride in raising certain fine chickens." During the show, Mrs. Knox's chickens won the Mitchell Cup, the show's award for the highest scoring pen of birds, as well as prizes for the best pen of solid color birds, the best pen of American class birds, and the best pen of White Plymouth Rocks. The *Enterprise* concluded that Mrs. Knox's chickens showed "that fine poultry can be raised right here in East Texas as good as any other part of the world." Grace may have developed her interest in poultry from her mother-in-law, Mary Knox, who also raised chickens in the backyard of her Dallas home. Mary continued the hobby into her nineties.

Grace Eva's death wasn't the only unpleasant event experienced in Texas by her mother, Eva Grace Fraser Briggs, who had been widowed by the death of her husband, Carlton Wilberton Briggs in 1880. In July of 1915—about seven years after Grace Eva's death in Polk County—her mother lived briefly in Galveston, Texas, when a flood devastated the island city. She wrote that she was among the last to cross the causeway to the mainland before the approach at both ends washed away. "I motored with relatives to Dallas and lived there six months," she said. In September of 1919, she undertook a ship voyage from New York to Galveston. "We reached Key West, Florida, a few hours after it had been struck by a typhoon...we met the aftermath of the storm, but missed by less than an hour being in the teeth of it," she said. Galveston, according to Mrs. Briggs, suffered little from the storm. "My friends had left the city (and) a colored servant met me and sent me to Houston," she said. In July of 1927, Mrs. Briggs, a resident of Milwaukee, Wisconsin, moved to Dallas, which she called "my future home."

On Saturday night, March 14, 1911—about a year after Hiram married Lillian Marshall, the lumber company's secretary and accountant—young Willie Knox, by now seventeen, was visiting in a neighbor's home in Knoxville when he was stabbed to death. An employee of the Knox railroad, John Frisby, 24, who had been seen with Willie in a room a few minutes before the death, was charged with murder although there were no eyewitnesses. Willie was buried beside his mother's remains in Forest Hill Cemetery on Sunday afternoon, March 15.

Frisby was released by Justice of the Peace M.S. Tew on a bond of $750 and indicted by a Polk County grand jury in the killing. The trial, however, ended in a hung jury two years later. J.L. Manry, then district attorney and later district judge, was related to Frisby and a special prosecutor was appointed. The case was continued several times and finally, in January of 1919, District Judge Manry was disqualified because of his kinship to the suspect. Attorney George Bean was appointed a special judge in the case, but the case was dismissed on a motion by prosecutors, who were unable to summon sufficient witnesses for a second trial.

Willie had been educated at a private school in Dallas and at Peacock Military Academy in San Antonio, where he was awarded a medal for heroism in extinguishing a fire in an academy building unassisted. Had he not gone for help, the fire would have gotten out of control, said Academy officials. A family biographer said Willie "was a lad endowed with intellectual powers above the average boy. His powers of observation were early developed and, at the age of eight, he had acquired a fund of general and useful knowledge. At seven, he read books much beyond his years and reason with the faculty of one twice his age."

On Saturday morning, June 28, 1913, Colonel Knox joined Willie and his mother in Forest Hill. On a business trip in Houston, the 78-year-old lumber tycoon was stricken by a heart attack while staying at the Brazos Hotel. He died at 7:10 p.m. Thursday, June 26. He had apparently been ill on and off for several years, but his body bore "a livid bruise across the face," reported *Master Detective* in a January, 1938, article. While Colonel Knox made his home in Dallas, Hiram and Lillian decided to bury him in the Livingston cemetery beside his daughter-in-law and his first grandchild.

Polk County would not forget the three Knox deaths. Years later, former Knox employees would connect the deaths with other tragic deaths in the Knox family.

One former Knoxville resident claimed in a 1987 interview with a Polk County historian that young Willie was stabbed to death by a man hired by "the Knox woman," although there were no such charges or evidence to support the claim. The Knoxville resident said the Knox story constituted "one of the biggest scandals in this countryside then, one of the awfulest messes that ever took place in Polk County." Another former Knoxville resident claimed that Grace Briggs Knox "died of the mumps; doctors back then were quacks and she had the mumps and they let her throat swell up and she suffocated." A third former resident said "some folks said it (Willie's death) was over insurance on him." None of the claims have ever been proven and may have been no more than rumors stirred by Hiram Knox's death by gunshot in 1922.

At the turn of a new century in 2000, little was left of the Knox sawmill in Polk County. Time had obliterated all physical vestiges of the mill and the Livingston and Southeastern Railroad, except a few slabs of concrete and graves containing the remains of Knoxville residents—such as the Creels, Malones, Blairs, Walkers, McKims and Parkers—in Bluff Creek Cemetery near Soda.

The sawmill's old mill pond, where logs were dumped and kept wet to preserve their freshness before being converted into lumber, is located in the woodlands near the Soda Loop Road. Some of the earthern remnants of old Knox railroad bed can be seen on the property line separating the Soda homesteads of Robert Upton and his neighbor, former Livingston jeweler Jerome Thomason.

The only other reminders of the Knox presence in Polk County are old photographs taken in the early 1900s. One is an early image of a grim Colonel Knox wearing a suit, tie and a wide-brimmed hat, standing by a stack of lumber. Another is a scene from a December 31, 1908, picnic by a Livingston group described by the *Polk County Enterprise*: "On last Thursday, a crowd of young people from this city...went out on the L&SE Railroad to Knox to spend the day...after dinner was over, Mr. (Hiram) Knox had the train carry the party to the woods and here we were shown the steam loader at work loading logs. Mr. Knox had gotten Mr. U. Moore, the photographer, to go along for the purpose of making some views..."

While Colonel Knox, Grace and Willie were seemingly in eternal rest at Forest Hill Cemetery, events that would eventually change as Lillian Knox's life continued to take strange twists and turns.

Willie Knox, victim of a 1911 murder at Knoxville in Polk County, Texas. *Photo courtesy, Merton Knox family.*

Around 1917, Lillian Knox posed for this photo, apparently at East Mayfield, Texas. *Photo courtesy, Texas Forestry Museum.*

The graves of three members of the Knox family were removed from Forest Hill Cemetery in Livingston in 1924. *Photo, Bob Bowman.*

5

EAST MAYFIELD. With Colonel William Hiram (Will) Knox—the family's contentious but wealthy patriarch—at rest in Livingston's Forest Hill Cemetery and work underway on a new Knox sawmill near Hemphill, Hiram and Lillian Knox prepared in 1913 to oversee its completion.

As early as 1900, Colonel Knox and Hiram had purchased 25,000 acres of virgin pine timberland from seven landowners, H.C. Maund, J.O. Davidson, W.S Arthur, W.G. Ener, G.W. Speights, A.D. Bennett and Alex Gilmer. The lands remained largely unlogged because the closest Knox mill was nearly ninety miles away in Polk County, but the Knoxes sold turpentine gathering rights in the pine forests to naval stores industries operating turpentine camps in Texas and Louisiana.

Located in the easternmost area of Texas, where the Sabine River separates the state from Louisiana, the red, rolling pinelands of Sabine County form one of the oldest regions of Texas.

In the early 1700s, Louis Juchereau de St. Denis led three expeditions into Texas, one of which took him along the old San Antonio Road in northern Sabine County. The road later became the main route of travel into and across Texas. After St. Denis, the area slowly began to settle. Original land grants from Spain and Mexico covered 220,000 acres. The largest, made to Juan Ignacio Pifermo in 1794, covered 17,713 acres near what is now the community of Geneva.

Stephen F. Austin, "the father of Texas," came through the county in 1820 on his way to colonize Texas, crossing the Sabine River on a ferry owned by James Gaines. He was followed in time by Sam Houston and other revolutionaries whose resentment against Mexico's oppressive government led to the creation of the Republic of Texas in 1836. Three Sabine County residents—Gaines, John S. Roberts and William Clark, Jr.—signed the Texas Declaration of Independence.

Following the Texas revolution, Sabine County was established in 1837 as one of the Republic's first counties, covering 546 square miles. It has the unique distinction of having the same boundaries today as it did then.

By 1920, the county remained largely a rural area with an estimated population of 12,299. The census that year showed the county had 1,270 farms; the principal crops were cotton, corn and sweet potatoes.

The county's major towns—Hemphill, Pineland, Milam and Bronson—were largely dependent on lumbering and farming. Hemphill has been the county seat since 1858 when Sabine County voters decided that Milam, the original county seat, was not centrally located. The courthouse was moved to the county's geographic center and a new town—named for John

Hemphill, a former Texas Supreme Court Justice and United States Senator—was created on eighty acres donated by Richard and Annie Slaughter.

In the 1920 census, the county had eighteen manufacturing establishments. Most were sawmills, which had begun to harvest the vast tracts of pine timberland stretching from one end of the county to the other. The sawmills' owners were anxious to fell the pine forests to produce lumber for sale in the growing cities of Texas, Louisiana and other southern states.

In 1911, Colonel Knox came to the Hemphill area and began scouting for sawmill sites. In his book, *Historical Sketches of Sabine County*, Cecil McDaniel wrote that Brad Carter of Bronson recalled that he sold Colonel Knox 100 tomatoes at a dollar apiece on a Saturday morning. A.V. White, another Bronson resident, said that while Knox was in the community, apparently looking at a potential railroad site from Hemphill to Bronson, he carried an umbrella. When two Mexican men, believing he could not speak Spanish, cursed Knox roundly, he struck both with his umbrella. The men apparently did not know Knox owned a lumberyard in Mission, near the Texas-Mexico border, and understood at least some Spanish, especially expletives.

Early Sabine County residents who met Colonel Knox in the period between 1911 and 1913 often told tales of his hard business deals and his anger. "He was the type who dominated all around him, if not by personality, then by violence," said the *Houston Press.*

Colonel Knox picked for his sawmill site a tract of land west of Hemphill, the county seat of Sabine County. The site was named East Mayfield for Texas Railroad Commissioner Earle B. Mayfield, who also served in the U.S. Senate from Texas in the

1920s. Knox and his son had apparently met Mayfield, a native East Texan, while planning their railroad and had courted the official's favor as they did with public officials in Wisconsin. Mayfield won his U.S. Senate seat in the 1920s, despite his identification as the Klu Klux Klan candidate in a campaign against Charles A. Culbertson, the incumbent Senator. Mayfield avoided the Klan issue in the campaign and spoke out against federal control of railroad freight rates, a position which probably appealed to the Knox family.

Although he lacked his father's drive and determination, Hiram Knox was no novice in the sawmill industry. He had learned the business well from his father and had become a competent inventor of sawmill and logging machinery. One of his inventions was a log loader which stood on its legs above railroad flatcars. The loader began loading cars at the front of the train behind the locomotive and moved down the train until all cars were filled with logs. The loader could then be transported to the next logging job by simply lifting its legs and loading itself on the last car.

In the early 1900s, logging and sawmill jobs were hard, dangerous and poorly compensated. The promise of new wealth had enticed northern capital into East Texas and jobs were open to anyone unafraid of work. The mills were operating at top speed, feeding trees hundreds of years old into the powerful saws of sawmills, providing lumber for homes and businesses.

Because the gypsy lumbermen frequently moved their sawmills from location to location, the homes and community amenities for their employees were poorly built, usually thrown together from reject boards from the mills. The clapboard houses stood beside dusty or muddy streets, depending on the weather, but they at least had four walls and a roof.

Wages were meager. Unskilled workers were paid little more than a dollar and fifty cents for a day that could last as long as ten hours. Skilled hands such as sawfilers and machine operators earned about three dollars.

Injuries were rampant in sawmill towns. Men lost hands, toes and legs while working in the mill and sustained broken arms and legs in the woods, which could be equally as dangerous.

Most sawmill towns had few amenities—perhaps a general store or a commissary operated by the lumber company, a school, a church and sometimes a community garden. Prodded by Lillian's concerns and her experience in dealing with poverty in Comanche County, she and Hiram set out to make East Mayfield more than an ordinary, run-of-the-mill sawmill town. They had likely visited places like Diboll, Pineland and Kirbyville, where progressive lumbermen like Thomas Lewis Latane Temple and John Henry Kirby had built permanent communities to enhance their modern sawmills. East Mayfield, Lillian argued, should not be a "cut-out and get-out" settlement.

Under Hiram's supervision, the sawmill at East Mayfield became one of the most modern in East Texas. It was a modern, double-band rig with dry kilns, planer and other auxiliary equipment and buildings necessary for producing 90,000 to 100,000 board feet of lumber a day for the growing marketplace. Together, Hiram and Lillian pushed the sawmill to completion. It began operations in 1915 with more than 400 workers. Knox encountered financial problems during the construction of the mill and was forced to sell 100 million board feet of shortleaf pine stumpage to North Texas Lumber Company, based in Anderson and Cherokee Counties, in 1915. The new mill finally cut its first log in 1917.

While the sawmill was being planned in 1912, Lillian suggested that the lumber company build a railroad from East Mayfield to the mainline tracks eleven miles away at Bronson. While Colonel Knox liked the idea, Hiram was cool to the idea. "How can we grow without a railroad?" she argued. "We're eleven miles from a railroad." Hiram answered: "Go ahead and build it yourself."

Lillian did. The Lufkin, Hemphill and Gulf Railroad (the LH&G) was chartered October 14, 1912, to build a line from Kindred in San Augustine County and east to Godwin's Ferry on the Sabine River, a distance of about thirty-eight miles. The principal incorporators were Hiram Knox of Hemphill, W.M. Caddy of Keenan, and Chester B. Collins of Lufkin. By 1913, nine miles of track from Bronson toward East Mayfield had been completed by Knox Construction Company, which had been contracted to build the entire line at cost, plus ten percent.

The road to East Mayfield was finished in 1916, a distance of 12.3 miles from Bronson, where it connected with the Gulf, Colorado and Sante Fe line at Bronson. Although Lillian was ambitious, no other expansions were undertaken and the railroad never touched Lufkin or the Gulf. It did, however, have a spur into Hemphill. In 1916 a photographer captured an image of a group of East Mayfield children who were being transported to Hemphill for classes. One of the little girls in the photo, Sibyl White, became the principal of Hemphill High School in the 1940s and was perhaps the first woman in Sabine County to earn a PhD (from Rice University, Houston).

Hemphill's wags soon christened the railroad as the "Left Hemphill and Gone," the "Lillian, Hiram and Gussie (Gussie was the Knoxes' household cook)," and the "Leave Here and Go." While Lillian supervised the railroad's operations, her secretary, T.C. Carson, was the traffic manager.

With the sawmill as its economic base, East Mayfield became "a model community," according to Robert S. Maxwell and Robert D. Baker, authors of *Sawdust Empire, the Texas Lumber Industry, 1830-1940.*

The emerging town, which was incorporated in 1914, had more amenities and employee conveniences than most sawmill towns its size in East Texas. Its employees' homes, numbering about forty, were larger and more comfortable. The town had several churches, a school, movie theater, a small hospital and a community center. The Knox offices, hotel and most of its business buildings were located on Italy Road, East Mayfield's main street. The Afro-American community was located near the dam on the mill pond road. Devereaux Chapman of Lufkin, whose parents lived in East Mayfield, was born in the town's community center after it was converted into apartments for Hemphill residents. "I heard many stories from my mother about East Mayfield and all the things it had," he said.

Lillian and Hiram built a large commissary store and issued metal and wooden tokens, which could be used in trade at the store or, at intervals, redeemed as currency—a common practice among sawmill towns of the era, but one which had been abhorred by Hiram's father in the Wisconsin white pine country in the 1890s. The East Mayfield metal coins consisted of a zinc center emblazoned with the currency amount, surrounded by a copper or brass ring shaped either in a circle or hexagon bearing the name "Knox Lumber Company." Former Knox employees said they were able to spend the sawmill money at stores in Beaumont and in other communities. The stores would then return the coins to the lumber company in exchange for U.S. currency.

In a 1956 interview with Robert S. Maxwell, John W. Minton, a Hemphill attorney who represented Knox Lumber Company, described the logging operations used to bring fallen trees to the sawmill: "The early operations used oxen to drag the logs to the tram with log wagons. This was slow, hard work. Most of the time the drivers owned their own oxen, but in some cases the company furnished them. Later, horses and mules were used, which were more manageable and faster, but less powerful. An innovation of the post-war years was the use of skidders. These were large machines with a long steel cable which could be unwound in the direction of the last tree cut and fastened to a large log and dragged to the tram road. This method was fast and economical, but destroyed all of the small growth and seedlings in its path. A great many miles of logging track were laid into the forests and the logs were carried or dragged to the railside. When that particular area was cut out, the tracks would be moved and laid down in a new place."

Minton said "front camps" were established in the woods by the tracks. Workers in these camps lived in shacks or housing quarters fitted inside railroad boxcars. Only men usually lived in the camps. Their supplies and food were brought in by rail on a daily basis from the sawmill town. "The wages were low and the hours were long by the logging crews," said Minton. "The food was usually quite coarse, but plentiful. There was no recreation, except what the men provided themselves. Usually, after a hard day's work in the woods, the men were too tired for any recreation. Both blacks and whites made up the logging crews," he said. Unions did not exist in the lumber business during the time the

Knoxes controlled East Mayfield.

The sawmill was just as hard and dangerous as the woods. To saw logs into boards, a man was strapped into a seat on the steam-powered saw carriage, which would make four quick passes down each log, sawing off slabs with tree bark. The remaining block of pine was then sawed into boards with thicknesses of an inch, two inches, four inches or more, depending on the ultimate use of the lumber. The lumber was then sent down a sloping conveyor, carried to a planer for finishing the surface of the boards, and then directed to a drying kiln. The boards were stored on the mill's yard until they were shipped to customers.

On Christmas Day, 1918, Lillian, only 27 years old, remained an attractive, slim woman, according to photos in a Knox family album. But a year later, she looked much older and considerably heavier. Hiram also looked older, fatter and bore a striking resemblance to his father. Suspenders, a tall black hat and a cigar in the corner of his mouth had become a part of his persona. In a reminder of a earlier photo taken of his father with his pet black bulldog, a hound dog stood at Hiram's feet.

Before 1920, the wealth of Lillian and Hiram Knox was estimated to be in excess of seven million dollars, made up largely of the East Mayfield sawmill, its supporting timberlands, the railroad, other property in East Mayfield, oil leases and various other properties Hiram had inherited from Colonel Knox. From the colonel's initial ownership of 10,000 acres in 1900, Lillian and Hiram had expanded the holdings to some 50,000 acres—making them one of the largest landowners in East Texas.

For his family, Hiram Knox built a large, rambling home—often called "the mansion" by mill hands. The size of the Knox home eclipsed that of houses occupied by nearby Hemphill's social gentry, including the Tooles, Pratts and Hamiltons, leading to resentment by some of the town's leading families. The house was described by the *Houston Press* on January 6, 1923, as "a baronial castle." The *Press* added: "The woman who might have lived in the luxury of a Riverside Drive (in Houston), if she had so desired, was content to build her fairyland in the lumber camp." The house was gradually expanded to include 22 rooms.

The home was soon occupied by Hiram and Lillian, along with six children, Hiram William Ambrose (known simply as Hiram), born August 17, 1911; William Hiram (Bill), born June 16, 1912; L. Marcene (Mark), born January 29, 1917; twins Sidney Merton Blackshear (Buddy or Bud) and Mary Lyllian, born October 6, 1918; and Mary Marguerite Signoria, born April 4, 1920. When the U.S. Census was taken in 1920, the report for the Knox household also included Clara L. Marshall 15, and Daisy E. Marshall, 12, Mrs. Gertrude Woods, 24, a maid, and Agatha W. Knox, 16, a servant.

The Knox family Bible, which was in the possession of the oldest Knox child, Hiram, when he died in 2000, showed that nine children, including three sets of twins, were born to Hiram and Lillian. When William Hiram was born in 1912, he had a twin, John, who died in infancy on September 12, 1912. Two other twins, Grace and Lillian, were born March 31, 1913, and died on October 31, 1914.

Because of Lillian and Hiram's tendency to repeat their own names, as well as those of Hiram's mother and father, there was considerable confusion among people coming into contact with the family. Within the family, there were three Hirams, three Lillians (one spelled Lyllian), two Williams, two Marys, and two Marguerites (Lillian's middle name was Marguerite).

Before he died in 2000, son Hiram Knox said Lillian left daughter Mary Lyllian with a Catholic convent in Shreveport, Louisiana, at the age of eighteen months. Hiram also said his mother often tried to give away her children or place them with relatives in other states. He said Lillian tried to place his brother Buddy with a man in Whitefish, Wisconsin, at the age of 12, but the man declined the offer, and Buddy was returned home. Lillian later told him the man was an uncle, he said.

The Knox children were often in the care of Elizabeth Kigle, the family's nanny, and her husband, Richard. The children often stayed in the Kigle's home, enjoying the company of the couple's two children.

As the oldest Knox boys, Hiram and William grew up in Hemphill where they were known for their pranks. During a Halloween week, the Knox brothers and several friends disassembled a small wagon and rounded up several animals and chickens during the night and led them up the stairs to a widow's walk atop the Sabine County Courthouse, fastened them to the structure, reassembled the wagon, and left the animals with food and water. Hemphill's townspeople were stunned the next morning.

During another Halloween observance, they caught a Hemphill man using his outhouse, lassoed the structure, tied the rope to a horse, and pulled the man and his two-hole privy a fourth-mile down a country road. "He cussed all the way," said Hiram.

When he was only five, young Hiram contracted typhoid fever, which left him with a crossed eye. He was fitted with glasses and when he was in his teens, he underwent three operations, which improved his sight only slightly. He also had mumps, which left him unable to produce children.

The Knox children often spent weeks visiting with Hiram's relatives in Wisconsin and with Lillian's sisters at homes Lillian and Hiram rented on Orange and Calder streets in Beaumont. They also spent summer months in Colorado Springs and Galveston. While staying at a Galveston hotel, Hiram and his brothers were admonished by the hotel manager for their conduct. To retaliate, they bought five pounds of Limburger cheese and spread it on the columns in the hotel lobby. "It sure made for a sweet smelling place," he laughed.

While staying at the Brazos Hotel in Houston, the Knox brothers rode their small pedal-driven cars from the hotel across a busy Houston street to the Southern Pacific railroad depot and went inside its restaurant for a sandwich "The police saw us, got a hold of momma, and she sent over a bellhop to escort us back to the hotel. We were grounded for a week," said Hiram.

The Knoxes traveled widely, which necessitated dependable automobiles. "We had the best. We never bought a second-hand car, " said Hiram. "The cars I remember best were a gray Cadillac and a 1916 silver Pierce Arrow," he said.

The Knox children also had the best toy cars money could buy. Photos in a family album depict the children playing with three miniature vehicles shaped like a World War I German ambulance, an American roadster, and a railroad engine. The ambulance carried a symbol resembling a Nazi swastika, but the symbol was reversed. Some of the toy cars enjoyed by the Knox children were motorized.

Lillian also spent some of her new wealth on her own family, the Marshalls. Her sisters and brother were frequent visitors to the Knox mansion, enjoyed vacations together, and received lavish gifts. In Texarkana, Lillian hired a cemetery architect to design a tall, angel-topped monument and a low fence of fine marble for the Woodlawn Cemetery plot of her mother, Daisy Marshall, who died in 1910. Standing to the right of the cemetery's entrance, the family plot was the most elaborate in the graveyard, the final resting place for many of Texarkana's wealthiest families. The monument bore the inscription: "She was the sunshine of our home." Buried in the plot with Daisy were two infant children, George Edwards Marshall, who died in 1901, and Iola Gertrude Marshall, who died in 1905.

To furnish her new home in East Mayfield, Lillian shopped for the best furniture, silverware, china and tapestries. One room in the home was devoted to a collection of fine paintings, including some by the old masters. Exquisite bric-a-brac had come from Lillian and Hiram's European honeymoon and from two trips to France, where she learned to speak French.

Surrounding the home was a fenced park stocked with deer, peacocks, birds of paradise, ducks and other forms of wildlife. The creation of an elaborate landscaped garden inside the park, supported by a large greenhouse, was supervised by a noted landscape artist. Hemphill residents knew the area as "the deer park."

Skilled artisans were imported into Sabine County to make the Knox mansion one of the finest in Texas. The house included innovations such as a complete sewer system, an electric lighting plant, and its own telephone exchange. Even after it was built, Lillian kept making additions to the home. "It grew like topsy, this way and that way," said son Hiram.

Hemphill's elite seethed as Lillian's home became larger and larger. But it wasn't difficult for Lillian's wealth to elevate her status in Sabine County. East Mayfield's closest neighbor,

Hemphill, was a poor, rural community best known for being the county seat. The town had two weekly newspapers and two fraternal groups, a Masonic lodge and an Order of the Eastern Star chapter. The town also had a number of churches, predominately Baptist, although it had neither a public library, a park, or many of the other niceties and amenities found in other East Texas communities.

But Lillian had plans to change all of that with Hiram Knox's sawmill and his inheritance.

Brothers Hiram, Marcene and William playing in a toy car in front of the Knox home at East Mayfield, Texas, in 1918. *Photo courtesy, Merton Knox family.*

Brothers Hiram, Marcene and William Knox were posed for this photo taken at a Knox Lumber Company logging scene near Hemphill in 1919. *Photo courtesy, Merton Knox family.*

A photographer caught the five Knox children dressed in their best clothes. Shown are Hiram, Marcene, William and twins Merton and Mary Lyllian. *Photo courtesy, Merton Knox family.*

Lillian Knox after returning from a trip to Austin, where she was elected chairman of the 14th Senatorial District's Women's Suffrage Movement. *Photo courtesy, Merton Knox family.*

The Hemphill, Lufkin and Gulf Railroad ran from Hemphill to Bronson, Texas. *Photo courtesy, Cecil McDaniel.*

Colonel William Hiram (Will) Knox, standing on a railroad track near Bronson before his death in 1913. *Photo courtesy, Cecil McDaniel.*

The office of Knox Lumber Company at East Mayfield, Texas. *Photo courtesy, Cecil McDaniel.*

The sawmill of Knox Lumber Company at East Mayfield, Texas, around 1920. *Photo courtesy, Cecil McDaniel.*

6

LADY BOUNTIFUL EMERGES. As one of Texas' wealthiest women, Lillian Knox was determined to make East Mayfield—a small, rugged sawmill community on the edge of the Texas-Louisiana border—the center of East Texas' social life.

Her mansion was a frequent site for lavish parties. Prominent couples from Lufkin, Nacogdoches, San Augustine, Shreveport and Beaumont were invited to join Lillian and Hiram. The men were captivated by Lillian's youth and charm, but their wives resented their hostess. If there were not enough attractive women in Sabine County to satisfy her guests, Lillian was likely to recruit young girls from the community, dress them in expensive formal dresses purchased from an exclusive Beaumont store, and encourage them to mix with her guests.

One celebrated event was held annually on June 26 to mark the anniversary of Colonel Knox's death. The Knox ballroom was adorned with a lifesized painting of the colonel.

Music, dining and dancing went on for three days. Limousines were dispatched to distant cities in Texas and Louisiana to bring guests to East Mayfield. As many as three orchestras were imported to keep the party moving. Elaborate decorations were provided by interior designers. It was said that Colonel Knox had asked that his friends and family enjoy themselves with an annual party on the anniversary of his death. While it seemed unusual that Colonel Knox's death was celebrated with music, revelry and food, few of the guests apparently complained.

Hiram, however, appeared bored with his wife's social functions. He was often found asleep on a couch as Lillian laughed and joked with her guests. An avid outdoorsman, he was happier tramping through the woods with his men or watching company loggers clear a new tract of pine timber. The *Houston Post* reported on January 6, 1923: "Hiram wasn't a chip off the old block. He cared nothing for business. They tell in Hemphill how the friends of Hiram feared he would die as an auto speed fiend. He was distant and had few friends. But on roads around Hemphill he could be seen almost any day in a high speed car breaking all speed records, usually going to some isolated spot with his dogs and guns on a hunting expedition. The only cronies he had were a few old hunters, or odd characters of the timber lands, and he cared nothing for the gay crowd upon which his wife was lavishing his fortune."

While Lillian spent large sums of money on her social friends, she was just as generous with the employees of Knox Lumber Company, where she directed the labor policy of the company. High wages were paid and a costly housing program provided homes for employees with reasonable rents. The town's finest homes were occupied by what one resident called "the supremes," the mill's superintendent and other key officials. These homes were located on the road leading from East Mayfield to Hemphill.

In the community, Lillian was a sainted figure. Everyone who worked for Lillian's railroad was given a new Ford automobile. Every newborn child of a Knox employee received either an expensive baby gift or a bank account and the parents received money if Lillian felt they needed it. Every Friday, if the weather was good, Lillian sat beneath a tent near her home and handed out cash to anyone with a story of need, even though Hiram and her friends cautioned her to slow down her spending.

She regularly visited the homes of the Knox employees, distributing money, providing health care and passing out gifts for children. Each Christmas she provided every mill family with a turkey and gifts.

Lillian was also an easy prey for a hard-luck story. When a piano salesman told her a pathetic story about his desperate need for a sales commission, she purchased three grand pianos. When she decided a young couple should hold their wedding reception in her home, she replaced every rug in the 22 rooms.

Lillian also spent lavishly on community causes. She donated significant sums to all churches in Hemphill, including a binge in which she paid all of the expenses incurred by a church in entertaining a local cleric. She also provided Hemphill's school with a modern playground, athletic equipment and showers. In a 1937 interview with the *Dallas Times-Herald*, she claimed she was instrumental in the founding of the Texas women's suffrage movement with a gift of $200,000.

One story involved a wedding in the Knox home. When a dozen of Hemphill's young men came for a rehearsal a few days before the wedding, they were without dress suits. Lillian made a quick call to a Beaumont clothier for the suits, which were rushed to East Mayfield with a tailor to make alterations. On another occasion, Lillian sent a locomotive to Bronson to bring to East Mayfield a special train carrying the influential members of the Beaumont Booster Club.

An early Sabine County resident, Robert Austin Gomer, wrote his niece Helen Schluter about World War I in Sabine County. He said "there were lots of patriotic meetings, rallies and the like. Mrs. Knox and some others from Hemphill came down to Sabinetown, organizing a Red Cross chapter." He said Mrs. Knox "was a social partier. She spent lots of money...I guess she had it to spend, too. She had the most expensive cars, I think, possibly a liveried chauffeur..."

But there was a darker side to Lillian. Those closest to her claim she had a dual personality. At times she would fly into violent rages and then subside into remorse and profuse apologies. Her servants feared her, but loved her. She demanded instant and unquestioning obedience from the staff. If it wasn't forthcoming, they faced dismissal and, in some cases, were struck by blows from Lillian's own hand. If she wanted the chauffeur to drive her to her second home in Beaumont, it mattered little if the roads were bad, the hour was late or the weather was poor.

Lillian's husband Hiram loved and feared her. He heard numerous rumors of extramarital affairs and soon grew to despise Lillian's dalliances and her paramours, but Lillian seemed oblivious. She seemingly knew she held a certain power over Hiram, as well as other men with whom she came into contact. Hiram was pliable and manageable in Lillian's view. When he was approached by mill employees on business matters, his reply was often: "See Lillian about that."

B.H. Windom observed that in one nine-day trip from Hemphill, Lillian attended a fashionable dance in Beaumont, gave birth to a daughter, made a shopping excursion to Houston, attended a reception with friends in Dallas, and returned by automobile to her home. When twins were born at 1:50 a.m on October 5, 1918, at the Brazos Hotel in Houston, she was engaged in business meetings three days later. Even though she fussed over her children, their care was entrusted to nurses and governesses at East Mayfield. When photos were taken of the Knox children, the nurses and governesses were frequently in the photos; Lillian and Hiram were seldom shown.

In 1917, when the United States entered World War I, President Woodrow Wilson asked a number of U.S. lumbermen to come to Washington and implored them to begin delivering large timbers to build warships. Lillian attended the meeting and when it was over, she called the president of the Texas division of the Sante Fe Railroad, ordered fifty flatcars, twenty-five furniture cars and a number of additional cars. She then put the Knox sawmill on a work schedule of seven days and seven nights. As the timbers came out of the mill, they were loaded immediately on the railroad cars and shipped to the shipyards. "I personally went to every junction's railroad dispatcher and tipped them to get our cars on the move," said Lillian.

As Lillian said later: "My husband was the boss (but) I did all the work. He was born a millionaire and hated work of any kind, but he was the finest of men."

As one of only two Texas mills able to saw timbers large enough for the ships, Knox Lumber Company's business began to boom as eighty-foot timbers were needed for the masts on sailing ships. Soon, the timbers were moving south to the Gulf Coast shipyards—each mast carried by two railroad cars because of its length.

But there was a downside. Each government order had an urgent delivery deadline with stiff penalties for late deliveries. With his laid-back approach to business, Hiram Knox was swamped. He couldn't produce enough large logs to meet his government contracts and began to miss the shipyards' deadlines and, as a result, incurred heavy penalties.

Lillian came to the rescue. She challenged Hiram to let her handle the logging deliveries while he managed the sawmill's operations. He agreed and Lillian pulled on her riding boots, mounted a horse and, equipped with a riding quirt, began to supervise the cutting of trees and moving the logs to the mill. If the men working in the woods didn't move fast enough to suit her, she wasn't hesitant to use language usually reserved for loggers. She was so successful that Hiram had to plead with her to slow down the log deliveries until the mill could catch up with the flow. The Knox mill was able to fulfill its contracts.

When a group of lumbermen organized a meeting to complain that the government's demands for war-time timbers would cause them to lose their regular customers, Lillian—the only woman in the group—was incensed. She told them: "Gentlemen, this is our country. Have you considered the possibilities of this war? You are putting your country on crutches."

As a result, Lillian made it her personal goal to spur deliveries of war-time timbers in the lumber industry. She developed lists of mills which had the greatest capacity to meet the government's needs and delivered the lists to the Department of War.

Texas newspapers began to lavish praise on Lillian for saving the lumber industry during the war. She toured Texas, with two of her sons, Hiram and William, dressed in tiny doughboy uniforms, and used the Knox influence to induce customers and fellow lumbermen to purchase bonds until she had personally sold nearly two million dollars worth. "We'd parade around in our little soldier suits, sit out in the hot sun, wriggling and twisting, and listen to her blabbering about war bonds," said Hiram. "We were just her dressed-up puppets," he complained

Every Knox employee was urged to subscribe to Liberty Bonds. Lillian also helped the Red Cross enlist 1,000 new volunteers from Sabine County and recruited nurses and aides for the 1918 influenza epidemic in the county.

No less than President Wilson awarded her citations for her war efforts. The Knoxes were praised and honored on the Eastern seaboard, and elegant private railroad cars were often seen parked on the railroad siding near the Knox home in East Mayfield. Lillian spearheaded an effort to raise money for building a ship commissioned for President Woodrow Wilson and donated the timbers for its construction.

In an article on the Knox family on January, 1938, *Master Detective* writer Margie Harris said, "Texans began to compare the Knoxes with the McKays, the Fisks and other Easterners who had achieved great wealth in the West. Eastern capitalists, Washington politicians and captains of industry were their familiars." The article said the family's wealth exceeded seventy million dollars at one time, but that was probably an exaggeration.

But the good times soon came to an end.

Gomer said the Knox mill "butchered a lot of virgin timber

in Sabine County." He also wrote: "When I was a kid, we used to have a wagon trail through virgin yellow pine from near Lemmie Tatum's place up on to the very edge of Hemphill. Of course, timber conservation was unheard of then...Knox Lumber Company logged the timber the fastest way they could to amass all the profit they could."

When the war ended, the demand for lumber began to decline. Sawmills began to close throughout Texas and the rest of the South. Land and lumber prices sagged. To complicate his financial position, Hiram Knox had purchased large tracts of timberland at high prices during the war years—lands that were worth a fraction of the prices after the war.

While the Knoxes were struggling with unaccustomed failure, other events were changing the world. The cost of living in Germany went up 73.7 percent. The U.S. Supreme Court upheld the women's suffrage amendment, which Lillian had supported. *Reader's Digest* was founded. Ford bought the Lincoln Motor Company for eight million dollars. King Tut's tomb was discovered. Model T Fords made their appearance. And Miriam (Ma) Ferguson was about to become the first woman governor of Texas.

Lillian explained the Knoxes' plight: "When the war ended, there were no customers for lumber. We were getting on our feet when the depression and the bank closings occurred and we had to sell. We were still millionaires when the government attached our bank accounts, our sales, and lands." She added, "The losses and reverses we had were caused by normal conditions, not by what we gave away."

The Knox estate left to Hiram and Lillian by Hiram's father in 1913 began to dwindle. Tract by tract, the family's timberlands were sold off to pay their debts. By 1921, Hiram was becoming increasingly moody and depressed, and his only companion was often his old hound dog, Tiger.

To recover from his financial reverses, he sold the East Mayfield sawmill, most of its supporting timberlands, and a part interest in the Lufkin, Hemphill and Gulf Railroad to Temple Lumber Company, which owned a competing sawmill at nearby Pineland. Southern Pine Lumber Company of Diboll was also owned by the Temple interests.

Arthur Temple, Jr., of Diboll said his grandfather, T.L.L. Temple, a deeply-religious and straight-laced businessman, knew he would have trouble striking a deal with the Knoxes, especially Lillian, so he turned to another member of the family known for his "hard living ways" to work out the acquisition. "He and Lillian had a lot in common, so they struck a deal," said Temple.

Located ten miles southwest of East Mayfield, Pineland was founded with a small sawmill and logging camp in 1902 on the Gulf, Beaumont and Great Northern Railway. The community was originally known as John Adams' Mill, but was renamed Pineland in 1904 when it was designated as a post office. In 1906, the Garrison Norton Lumber Company took over the town's lumbering operation and constructed a new mill a year later. In 1910, the company had decided to leave Pineland when Thomas Lewis Latane Temple, a part-owner who lived in Texarkana, purchased the plant and renamed it Temple Lumber Company.

The town was at first composed almost exclusively of sawmill workers and their families, but the town began to grow after Temple's acquisition. Temple built a large commissary store and issued tokens, which could be used in trade at the store or, at intervals, redeemed as currency.

Like Colonel Knox, the Temples had some roots in Texarkana, and they likely knew each other, since Texarkana was the lumber capital of Texas around the turn of the century. The *Texarkana City Directory* for 1899-1902 listed twelve lumber dealers or companies, despite a population of only 18,054—a good reason why the city called itself "the gateway to Texas."

Angelina County Lumber Company of Keltys, a community near Lufkin, also made a bid on the Knox properties and reached an agreement with Hiram Knox on two separate occasions. Lillian, however, vetoed each of the agreements, incurring the wrath of Ernest L. Kurth, one of the owners of the Keltys company and among Texas' most prominent lumbermen. Kurth openly admitted his dislike of Mrs. Knox. He said Mrs. Knox "looked like an angel, but she was a devil."

To purchase the Knox properties, Temple Lumber Company paid $1,519,811 in cash and notes. The transaction included the Knox sawmill and planing mill, dry kilns and dry sheds, lands and timber rights, all sawed lumber and stumpage, log cars and locomotives, log loaders, livestock, logging equipment, employee housing (which rented for six to ten dollars a month), blacksmith and machine shops, the railroad's round house, commissary, repair shop, power house, the Knox tram road and the LH&G Railroad, oil house and fuel car, a pump house, hotel and boarding house, office and other business buildings, a restaurant, churches, schools, the shipping office and loading dock, and the mill pond. The transaction also included land titles and timber rights to 74 tracts of land totaling 17,500 acres and the Knox mansion.

Temple workmen began immediately to remodel the Knox mills to conform with Temple's needs. One of the first renovations in 1921 was the installation of a lathe mill, the blueprints of which still survive at Lufkin Foundry and Machine Company (now Lufkin Industries, Inc.) in Lufkin.

After paying the outstanding debts and taxes, Hiram emerged from the sale to Temple with only $150,000—a pittance when compared with his earlier worth—and even that wasn't immediately collectible. The government had begun to harass him with threats of non-payment fines over significant claims for surtaxes. It was an embarrassing turn of events for a family which had done so much for the government's efforts during the war.

Although Hiram wasn't exactly impoverished—he still owned 10,000 acres of East Texas timberland, substantial oil leases, and land in the Rio Grande Valley and Mexico—his days of vast wealth were gone. But what bothered him most was the fact that he had sacrificed the estate of his father.

He was now 58, an old man in his own mind. His depression deepened and he closeted himself in his home. A cryptic note penned beside an album photo of Lillian and her children on a 1920 vacation in Colorado said: "Hiram not well."

A side view of the Knox mansion at East Mayfield in 1918. *Photo courtesy, Merton Knox family.*

Elizabeth Kiger, the nanny for the children of Hiram and Lillian Knox, with two of the children in a family automobile. *Photo courtesy, Merton Knox family.*

East Mayfield children prepare to board the Lufkin, Hemphill and Gulf train to attend school in Hemphill. *Photo courtesy, Cecil McDaniel.*

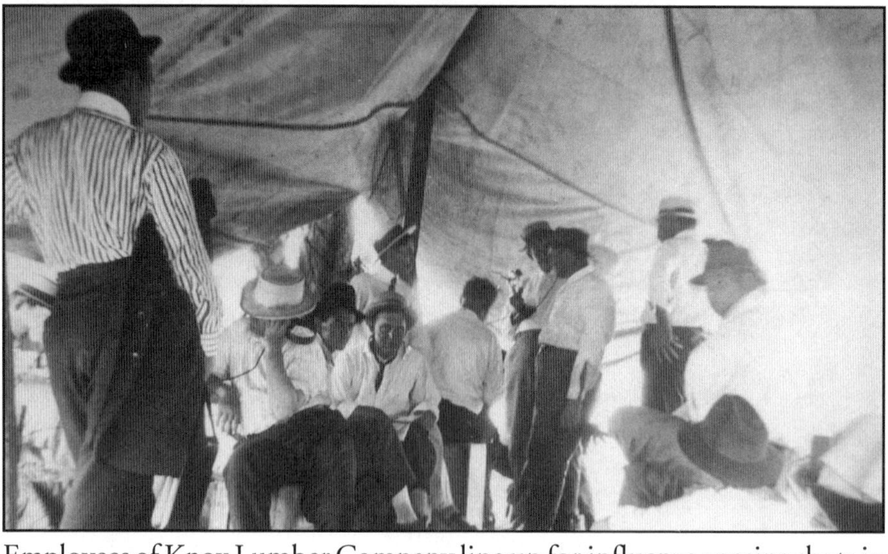

Employees of Knox Lumber Company line up for influenza vaccine shots in 1918 at East Mayfield. *Photo courtesy, Merton Knox family.*

Knox brothers Hiram and William were dressed up in doughboy costumes and accompanied their mother on a Texas campaign to sell Liberty Bonds during World War I. *Photo courtesy, Merton Knox family.*

Lillian Knox with twins Merton and Mary Lyllian shortly after their birth in 1918. *Photo courtesy, Merton Knox family.*

Lillian Knox placed this monument atop the grave of her mother, Daisy Marshall, in Texarkana's Woodlawn Cemetery. *Photo, Bob Bowman.*

7

DEATH BEFORE DAWN. Dr. William Thomas (Tom) Arnold, Jr., one of nine children of a Confederate soldier who lived in the Hemphill area, was among Sabine County's most respected citizens in 1922. In 1906 he earned his medical degree from the University of Tennessee and returned home to establish a medical practice on the East Texas frontier.

Known simply as "Doctor Tom" by Sabine County residents, Arnold influenced thousands of lives with his skills, compassion and devotion to medicine. Jim Lewis, a patient, said Arnold "did a lot with what he had" and possessed the natural ability to "make you feel like you were getting well and, sure enough, you did."

Dr. Tom's earnings were meager. His cash income seldom exceeded $4,000 a year and his services were often paid with fresh vegetables, pigs, chickens, stovewood, homemade preserves and syrup. He made house calls for two dollars and provided nine months of care and delivery for a mother and child for twenty-five dollars. He suffered from many of the frontier diseases which struck his patients.

When the Knox Lumber Company came to Hemphill in 1913, Dr. Arnold's practice improved considerably. Hiram Knox made him the company's doctor, as well as the Knox's family personal physician. In time, Dr. Arnold became a close friend of Knox, sometimes accompanying Knox on his frequent hunting trips, and even naming one of his children Hiram. Young Hiram Arnold eventually followed his father into medicine, becoming a gynecologist in Houston.

Working 24 hours a day, seven days a week, Dr. Tom battled diseases with the barest of medicines, mostly aspirin and morphine. Pneumonia was often treated by opening the patient's chest cavity to allow the pus to drain away. Surgeries and broken bones were repaired with only the application of "deadening medicine." When a Sabine County man's abdomen was sliced open by a knife-wielding drunk and his intestines fell on the ground, Dr. Tom washed them with distilled water, stuck them back into the man's stomach, and sewed him up. The man survived.

In the early morning hours of Sunday, November 26, 1922—slightly more than six months after her admirers had honored Lillian Knox as Lady Bountiful—Dr. Arnold was awakened by the ringing of his telephone—something not uncommon in his life.

The caller, Fannie Watson, Lillian Knox's sister, was hysterical. "Somebody shot Hiram in his bed. He looks like he's dead, " she shouted.

Arnold tried to calm Mrs. Watson. Asking a few questions, he told her he would come immediately to the Knox home. Before leaving, he called Sheriff George W. Alford, who was nearing the end of his term after losing a reelection bid to his chief deputy, E.W. Nations.

Arnold and the sheriff arrived at the Knox home almost simultaneously. They learned that a single gunshot had shattered the quiet of the residence north of Hemphill around 3 a.m. The family had earlier moved out of their landmark mansion in East Mayfield following the sale of Knox Lumber Company and all of its buildings, including the mansion, to Temple Lumber Company, and had moved into a home near the Padgett Town community, near a depot used by the Lufkin, Hemphill and Gulf Railroad.

The household servants were aroused by the anguished cries of Lillian, who slept in a room adjoining her husband's bedroom on the first floor of the Knox home. Knox was in his bed, lying on his left side, with a .45 caliber automatic resting loosely in his right hand. He appeared to be sleeping, but blood was oozing from the back of his head. A single empty shell lay nearby.

Dr. Arnold administered Lillian a sedative from an open bottle on Hiram's bedside table. Because of her condition, Arnold suggested that Sheriff Alford delay his questioning until she was calmer.

Alford began a cursory search of the bedroom, finding two typewritten letters addressed to Hiram's mother, Mary Knox of Dallas, and his attorney, E.J. Mantooth of Lufkin. The notes were discovered in a pocket of Hiram's coat, which hung nearby. Both had been typed on the stationery of the Brazos Hotel of Houston, an establishment Hiram and Lillian used on their frequent trips to Houston. The notes had been written a few days earlier.

Unable to question Lillian, Alford focused for the time being on Fannie Watson, the wife of a Houston lumberman. She said Hiram had retired early the night before while she and Lillian had sat up, talking until they, too, retired. She, Hiram and the servants were the only occupants of the house. The Knox children, she said, "were in the east attending school."

While two Texas newspapers reported that the Knoxes' oldest son, Hiram William Ambrose, 11, was sleeping in an adjoining room and was awakened by the shot, Hiram told members of his family decades later that his mother had sent him and the other Knox children to spend the week with relatives. Young Hiram, who was in Tennessee, returned home following the shooting, and recalled seeing his father "in his casket with a cork in his head." He said his brother Merton was in Wisconsin.

Mrs. Watson told the sheriff Hiram had been depressed, uncommunicative and spent much of his time alone in the house. Although the contents of the two notes were never disclosed, Alford apparently felt they were indicative that the lumberman had planned to commit suicide.

A day later, Alford began questioning Lillian. She confirmed Fannie's observation of Hiram's moody disposition. Even though the Knoxes had recently sold the East Mayfield sawmill and most of the family timberlands to Temple Lumber Company, she claimed Hiram had suffered significant losses in commodity markets and had incurred heavy government surtaxes. The family's financial condition was poor, she said. She also confirmed she had accompanied Hiram to Houston earlier in the week.

When officers questioned Hiram's associates in the lumber business, they also confirmed that he seemed depressed on several occasions. One quoted him as saying: "I was going to Mexico and start over, but now I know a better way out."

It seemed to be a simple case of suicide. But Alford and Dr. Arnold were not convinced.

Alford returned to the Knox home and began a more thorough search of the crime scene. He found that an open window in Knox's bathroom bore scuff marks as if someone had entered the window. Outside, on the ground beneath the window, he found partial shoe prints, which were unusually long and narrow. On the carpet near Knox's bed, Alford discovered fresh dirt matching the dirt outside. He also found a small bloodstain on the carpet. When Knox's body was examined by Dr. Arnold the next day, he could not find powder marks on Knox's head or hands.

Bothered by the circumstances, Sheriff Alford contacted the Houston Police Department, asking detectives there to check out the notes supposedly written at the Brazos Hotel. When the detectives reported back to the sheriff, they said they had found that the Knoxes had rented a typewriter while staying at the hotel and confirmed with the typewriter rental agency that the notes had been produced by its typewriter. They also questioned stenographers at the hotel, but found no one who had worked for Hiram in Houston.

To complicate matters, Hiram could not type, but Lillian had stayed with him in Houston. She was, it turned out, an able typist, having refined her skills as a secretary for the family's business before she married Hiram.

On Wednesday, November 29, Hiram Knox's body was carried by train to Texarkana, where his funeral was held at 2:30 p.m. at the Olive Street residence of Roy J. LaRue, a marble worker at Grigson Marble and Granite Works. Lillian had been married briefly to LaRue in 1906 and apparently remained a friend. Reverend Walter J. Jennings of Texarkana's Central Christian Church conducted the services and Hiram was buried in a family plot just inside the entrance to Woodlawn Cemetery, the town's most exclusive burial ground.

Hiram's pallbearers were T.L.L. Temple, Jr., Louis Heilbron, W.R. Hodge, H.V. Beasley, Tom Webber, and I. Schwarz. The *Daily Texarkanan* reported that those accompanying Hiram's body to Texarkana were Lillian; Mr. and Mrs. Percy Weiss, A.E. Brulin, Miss R. Ray, Miss Clara Marshall, Lillian's sister, Mrs. Alexander Marshall, a sister-in-law, and Mr. and Mrs. Felix McReynolds, all of Beaumont; Mr. and Mrs. J.L. Watson of Houston (Mrs. Watson was also Lillian's sister); C.B. Halbert and John W. Minton, both of Hemphill; and Mrs. Lewis Boggus of Mission.

Sometime after the funeral, Lillian and members of her family returned to Texarkana to plant near Hiram's grave a young pine tree, supposedly a symbol of the pinelands which produced the family's wealth. Four cedar trees were planted at the corners of the plot.

Sheriff Alford and Dr. Arnold, meanwhile, had concluded in Hemphill that Knox's death was not a suicide, but murder. They based their belief on a series of conclusions:

• Knox's hair had not been singed by the gunpowder. They said the gun would have to be held at least a foot from the head to avoid powder burns, which was impossible since Knox was shot in the rear of his head. The shot entered Knox's head about halfway between the right ear and the center of the back of his head. It lodged over the left eye.

• They cited the open and scuffed bathroom window, the

footprints outside and the scraps of dirt on the bedroom floor.

• They also produced documents showing Knox's financial condition was not as distressed as they were led to believe. Further, Hiram had deeded most of his property in Texas to Lillian and had made her the executor of his estate in documents executed over a span of two years. The paper trail was found in deeds executed or filed in Dallas County, Sabine County, Harris County and San Augustine County.

The most damaging evidence was that Hiram, at the time of his death, had $150,000 in cash from the sale of the Knox Lumber Company to Temple Lumber Company and had inherited $600,000 from an uncle, apparently Samuel Knox, in the eastern United States. The inheritance could have been transferred to Lillian in documents dated November 21, 1922, only five days before his death, deeding all of his lands and property to her. The November 21 deed was filed for record on November 22 in neighboring San Augustine County and wasn't filed in Sabine County until March 19, 1923, four months after Hiram was found dead.

Sheriff Alford, known for his doggedness in pursuing tough cases, kept plowing deeper in the case, helped by a team of private detectives hired by a friend of Hiram. A Houston newspaper said the detectives were reportedly employees of the Pinkerton Agency, but the agency declined to confirm the story.

Alford learned that the rumored rift in the Knox household was, indeed, factual. Servants told the sheriff of frequent arguments between Hiram and Lillian. Hiram often complained about his wife's excessive spending, especially for extravagant formal parties in a small town where she was already the community's leading socialite. He also complained about her propensity to hire expensive performing groups and artists from Houston, Dallas and Kansas City for performances and exhibitions in Hemphill and East Mayfield. Some of the most damaging comments came from John and Susie Hodo, servants inside the Knox household.

Hiram had threatened to change his will, revoke Lillian's power of attorney, and slow down her lavish spending. He felt Lillian was surrounded by people who only had an interest in the Knox money.

And then there was Chester Collins, a flamboyant Lufkin attorney who had been associated with the Knoxes for more than ten years. With a penchant for expensive suits, shoes and ties, Collins was regarded as one of the leading lawyers in Lufkin during the twenties and thirties and represented some of the leading businesses in East Texas. Collins also enjoyed a close relationship with Lillian, something Hiram knew and disliked. John Hodo told investigators that he had heard Hiram shout to Lillian: "If you don't keep Chester Collins away from here, I'll kill him."

Years later, Lillian told her attorney, Edward J. Barrett of Chicago, that Hiram knew of her supposed marriage with Collins. "He (Hiram) loved me as much as he could love any woman, but he did not want me to leave him...he told me before we were married that he needed a wife to protect him from women and blackmailers. I loved him for his goodness." It is not known if Knox also knew of Lillian's former marriage to Roy J. LaRue in Texarkana.

I.D. Arnold, the nineteen-year-old Knox chauffeur, told investigators that on the Monday before Hiram's death he had driven Lillian to the banks of the Attoyac River, where she met and talked at length with a man later identified as Chester Collins. Arnold said on the following day he had also driven Lillian to Beaumont to join Hiram, who was in the city for a business meeting, and later drove them to the Brazos Hotel in Houston. Arnold said when he went to the Knoxes' hotel suite, he saw Lillian operating a typewriter.

Incriminating evidence also came from another Knox employee, T.C. Carson, who served as Lillian's secretary. He said he had picked up a black mourning dress for Lillian at the railroad express office on the morning of Hiram's death. Investigation disclosed she had ordered it from Levy Brothers, a Houston department store two days earlier, telling the clerk it should be suitable for a funeral and should be delivered in Hemphill by Sunday. Carson also said Hiram had gambled heavily during the recent Houston trip and had brought home a five-gallon container of whiskey. Carson also said he bought, at Lillian's request, a bottle of the sedative found on Hiram's bedside table.

On November 27, a coroner's report issued in Hemphill declared Hiram Knox did not commit suicide. The death certificate said that Hiram died as the result of a gunshot wound to his head. Feeling he had enough evidence to lodge a murder charge against Lillian Knox, Sheriff Alford obtained on December 16, 1922, a warrant for the arrest of the woman everyone called "Lady Bountiful."

According to some accounts, Lillian spent the Christmas holidays with friends and family members in Beaumont. When she returned, Alford was waiting. On the morning of January 1, Alford called Lillian and drove to her home. He found his suspect dressed and waiting for him with a packed overnight bag. He drove her to the Sabine County Jail, located beside the County

Courthouse, placed her in a cell, and turned the key in the door. He then clipped his badge to the warrant, gave it to his chief deputy, E.W. Nations, and told him: "It's now your job and your case." Nations had defeated Alford in his reelection bid earlier in 1922, and it was likely that Alford felt he could extract some measure of revenge by turning over to Nations a controversial case of murder in which one of Nations' principal supporters was the suspect. Lillian Knox had written nineteen letters to her friends, urging them to vote for Nations.

Nations released Lillian from the jail cell and asked her to seat herself in his office while he called John W. Minton of Hemphill, the attorney for Knox Lumber Company. When he arrived, Minton found Mrs. Knox demanding that she be released on bail. Advised by Minton that murder was not a bailable offense in Texas, Lillian insisted he do something.

Conferring with a local justice of the peace the next day, Minton persuaded the official to release his client on bond. When the justice of peace agreed to free her on a $25,000 bond, Lillian told Minton: "Ridiculous. Five thousand will be sufficient." Within 24 hours of her arrest, she was released—the only accused murderer to leave a Texas jail before an examining trial. Lillian paid the $5,000 bond with a personal, uncertified check and returned to her home in East Mayfield after spending only one night in jail.

Another curious event occurred on January 2. W. Carter Arnold, who had recently retired as county judge in Sabine County, killed himself with a shotgun, igniting unconfirmed rumors he had been connected to the Knox death. Arnold, the father of Knox chauffeur I.D. Arnold, had been in poor health for some time. A portion of Judge Arnold's head had been blown away by the shotgun blast. His death was ruled a suicide.

On January 3, Hiram's mother, Mary Knox of Dallas, was contacted by *United Press* regarding her son's death. Her only response was: "It'll come out at the trial." Newspapers later reported that Mrs. Knox was financing the search for her son's killer, but it wasn't disclosed if she had hired the Pinkerton Agency.

On the same day, the *Houston Post* reported: "Some say that the district attorney (of Sabine County) has evidence of a well-defined plot to get rid of Knox. Just what part the widow is alleged to have played in this supposed conspiracy is not known." The newspaper also said that a statewide search was underway for a Beaumont man "who lavishly spent money and is said to have been connected with Mrs. Knox in several business deals...the man has figured prominently in sporting events in Beaumont and other South Texas cities." The man reportedly lived in splendor, spent money lavishly and gave huge tips. The press reports did not mention that Mrs. Knox's brother, Alexander (Alex) Marshall, lived in Beaumont and was involved in business deals with Lillian.

Lillian, meanwhile, took a jab at the people of Sabine County in an interview with the *Houston Post*: "Isn't it strange how they fawn at your feet and accept all that you have to give when the sun is shining, but when dark clouds come over your life, how so many of those fair weather friends flee and some of them, like wolves, are ready to devour you."

County Attorney T.B. Hamilton sparked another mini-mystery when he said that on the morning of Knox's death a large vehicle was seen speeding away from the Knox home, which was about 300 yards north of Hamilton's own home. Hamilton said he knew the make of the car and that it had been traced to San

Augustine, 25 miles away, but no further mention was made of the car.

Another press dispatch said the typewriter on which the supposed suicide notes had been written had been locked in a vault in B.M. Garrett's office in the Scanlan Building in Houston. The connection with Garrett was that the typewriter had been loaned by him to Charles Dickensheets, who stayed at the same Brazos Hotel suite occupied later that week by the Knoxes. Dickensheets was a well-known Texas newspaperman who made a fortune in oil and gas investments.

Meanwhile, John W. Minton, the Knoxes' Hemphill attorney, was assembling an expert team of lawyers to represent Lillian. Besides Minton, the team included experienced criminal attorney Steve King of Beaumont and Chester Collins of Lufkin. Collins, the son-in-law of E.J. Mantooth, had also represented the Knox business interests, but was also known for his criminal defense work.

Meanwhile, the news that Lillian Knox had been arrested for killing her husband sent shock waves throughout Sabine County and Texas. The leaders of the lumber industry were divided on her guilt. Such was also the case in Hemphill and Sabine County. Those who had benefited from her generosity defended her staunchly, but those who were envious of the Knox wealth were willing to condemn her.

Regardless of the diversity of opinion in Sabine County and Texas, everyone agreed without any dispute that the forthcoming trial would be the best attended event in Sabine County since Lady Bountiful's big appreciation event in June of 1922.

But there would be no trial.

Lillian Knox with her husband Hiram, right, and a friend on an excursion to Hot Springs, Arkansas. *Photo courtesy, Merton Knox family.*

Dr. Tom Arnold, the Knox family physician and a witness in Lillian Knox's preliminary hearing. *Photo courtesy, Cecil McDaniel.*

Sabine County Sheriff George Alford arrested Lillian Knox for murder on January 1, 1923. *Photo courtesy, Cecil McDaniel.*

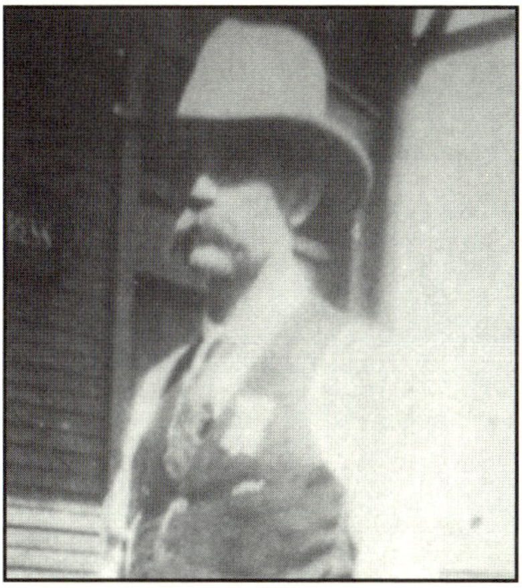

Deputy Sheriff E.W. Nations succeeded George Alford as sheriff of Sabine County on the day Lillian Knox was arrested for murder. *Photo courtesy, Cecil McDaniel.*

In a photograph taken in Texarkana, Lillian Knox, right, appears with two of her sisters in the Marshall family, which had roots in Texarkana. *Photo courtesy, Merton Knox family.*

Before his death in 1922, Hiram Knox looked much like his father, Colonel William Hiram (Will) Knox. Tiger, his hound dog, was a constant companion. *Photo courtesy, Merton Knox family.*

8

MURDER OR SUICIDE? Before Lillian Knox could be tried for Hiram Knox's murder, she faced two preliminary procedures.

One was a hearing on the charges before a local magistrate to determine what sum of bond should be fixed and if she should be handed over to a grand jury. The second was a review of the facts by a Sabine County grand jury.

Justice of the Peace W.H. Pratt, who also served as coroner in the case, announced to the press that he would make a full investigation to help him set the bond amount. Lillian's supporters, confident she would be acquitted, shrugged at Pratt's announcement. "Let him investigate," they said, "it's required by the law."

On January 3, 1923, two days after Lillian's arrest, she showed up at the Sabine County Courthouse, Hemphill's largest building. The brick courthouse, built in 1909 after a fire had destroyed Hemphill's original wooden courthouse, shared the courthouse square with a separate two-story brick jail.

Chatty and confident, Lillian wore an appropriate black dress and hat, the latter pulled down over her eyes. Escorted by her brother and sister—Alex Marshall of Beaumont and Fannie Watson of Houston— she went straight to the crowded courtroom. Alex shielded her from a phalanx of cameras with his bulky body.

Newspapers throughout Texas were fascinated that a prominent millionairess had been charged with the murder of her equally prominent husband. They converged on sleepy Hemphill from Houston, Dallas, Beaumont, Shreveport and other cites. They jostled with each other in the small courtroom to find the best place from which to view the widow and hear the conversation. And when the Knox headlines were written, they pushed world and national events off the front page.

At 9 a.m. Pratt opened the hearing and asked County Attorney T.B. Hamilton, the state's prosecutor, to make his opening statement. Hamilton said he would produce evidence that Mrs. Knox had caused the death of her husband on November 26, 1922, by shooting him, or having him shot, with a .45-caliber high-powered pistol loaded with a steel-nosed bullet. He said he would also produce evidence Hiram had been drugged with a powerful sleeping potion procured by his wife. He said the body was found in a sleeping position in his bed and because there were no powder burns on the body, he could not have shot himself. Further, he said the angle of the bullet's path through the victim's brain would prove he could not have shot himself.

Hamilton said the former sheriff of Sabine County, George Alford, would testify that he found an open window in the bathroom adjoining Knox's bedroom, the screen removed, and a man's footprints in the earth beneath the widow.

The prosecutor concluded by claiming the supposed suicide notes found in Knox's pockets were forgeries and had been typed by Lillian. Hiram, he said, did not know how to type. Hamilton also said the state had ample evidence of marital discord over business and personal matters, which he said led Lillian to cause Hiram's death.

Steve King, the lead attorney on Lillian's defense team, told the court he would not make an immediate statement, except to say "my client is innocent of the horrible crime and the facts will prove it." Privately, the defense team told reporters it would contend Hiram had "gone in for violent excesses," even purchasing a gambling house and saloon across the Rio Grande River near Mission, Texas, implying that someone from his darker side had killed him, not Lillian.

When testimony began, Dr. Tom Arnold was the lead witness. He described the events that occurred during the morning of November 26, when he was called to the Knox home. He said Knox was found "in an attitude of sleep" and that his eyes were closed.

Arnold's only new disclosure was that he had been called to the Knox home several days before the shooting to administer medical help to Mrs. Watson, Lillian's sister. He said Mrs. Watson made a statement to him at that time, but the defense raised an objection to Arnold's statement. The objection was sustained by Pratt.

Former sheriff Alford followed Dr. Arnold to the witness stand. His only new disclosure was that he had found two fresh footprints beneath the bathroom window of Knox's bedroom—both long and narrow—in the dirt. He said he found six cartridges in the magazine of the pistol and that only one shot had been fired.

He said there were no powder burns on Knox's head or hands.

When Katherine O'Keefe, a railroad express agent, testified that Lillian had received a black mourning dress before Knox's death, King heatedly objected to her testimony. "This woman has been tried all over East Texas by hearsay evidence, but the time has arrived when gossip ceases and nothing but the cold-blooded truth is wanted." Pratt, however, allowed O'Keefe's testimony to stand.

Sebe Clark, a Knox sawmill employee, testified that Mrs. Knox had recently told him she "would run the company or there would be trouble sure enough." He also said that Mrs. Knox had once told him of trouble with her husband and promised Clark she would give him a home if he would "help me out."

O.A. Crockett, the owner of a restaurant twenty miles south of Hemphill, testified that Mr. and Mrs. Knox had dinner at his establishment the Friday night before Hiram's death on Sunday. He said Knox had not been drinking.

T.C. Carson, Lillian's secretary, told the court Lillian had told him that Hiram purchased five gallon jugs of whiskey on a recent trip to Houston, that he had been gambling and lost heavily, and was drinking heavily at home. Some Hemphill residents who knew Knox doubted Carson's testimony; they said they had never seen Hiram drunk in all his life. "It don't sound like Hiram," said one Hemphill observer to a *Houston Press* reporter. Carson also said Mrs. Knox had asked him to buy a bottle of the sedative at a Hemphill drugstore, which he did. Drugstore employee E.H. Murphee confirmed the sale in testimony.

Mrs. Carson said said she had seen Knox's body shortly after his death and examined the bullet wound. She said the hair around the wound had been burned as though by powder. She was the only witness to testify about powder burns. Hemphill insurance agent R.H. Minton testified he had met with Lillian Knox on several occasions regarding insurance for her husband. He said she told him she wanted to purchase some insurance, but failed to do so.

Susie Hodo, a servant in the Knox home, said Knox had told her that he "was broke and couldn't stand it." She said Hiram had promised "to end it all." Susie also went into a detail of the case not yet explored. She said on the Thursday before Hiram's death, she received a long-distance call from Lillian, ordering her to retrieve certain company records from a cedar chest in the house and carry them to Chester Collins in Lufkin. She said she complied with the order. Susie also said she confirmed that the so-called "record house" at the sawmill, containing records of the Knox interests, had been destroyed by fire the day before Lillian issued her order.

John Hodo, Susie's husband who also worked in the Knox household, testified he had heard Knox say he "was going to keep Chester Collins away from my home or kill him." Hodo said Lillian asked him not to tell Hiram about the packages she was sending to Collins. The disclosures by the Hodos startled the courthouse. Lillian Knox's own attorney, who was defending her, was being implicated in the murder.

Following the first day of the courtroom proceedings, a reporter wrote: "Throughout the ordeal Wednesday, Mrs. Knox, Lady Bountiful of the countryside around Hemphill, smiled through her pallor as witness after witness gave what appeared to be damaging testimony against her. At times, she chatted animatedly with relatives and counsel. Alex Marshall, her brother, remained with his sister throughout the day, guarding her from reporters and photographers."

But Lillian wasn't overly concerned about being shielded from the press. That evening, she conducted a press conference at her home. She entertained the reporters lavishly and read the following to them: "I have no fear of the outcome. I am absolutely innocent of this terrible thing. I want my friends to know that I am guiltless, and when the trial is over, I will have convinced them of the fact beyond any doubt." She followed her statement by leading the reporters to the nursery, where her five children, now back home from a stay in the east, lay sleeping. "This is the little songbird," she said, picking up the smallest child, three-year-old Mary Marguerite, and hugging her to her breast.

She then led the press representatives through the room where Hiram was found dead. "This is the room," she said, and then corrected herself, "Yes, this was our room." She paused and gazed for a moment at Hiram's bed. Tears came to her eyes. "It's been so lonely, so lonely here since that awful night," she said.

Her story made front-page news throughout Texas with large headlines and photographs of Lillian, her former mansion, the sawmill and other scenes, but few photos of Hiram. One of the reporters wrote that Lillian "was a woman of noticeable personality, agreeable, quick-witted and gracious to all."

Next day, as the hearing resumed, prosecutor Hamilton recalled former sheriff Alford and qualified him as a pistol expert. Alford said he had test-fired the gun he found in Knox's bedroom and found that it had caused powder burns on sheepskin when fired three to six inches from the skin. This, he said, led him to conclude that Knox had not shot himself. He said there were no powder burns on Knox' head.

When he was summoned again by the prosecution, Dr.

Arnold said it was unlikely a man could shoot himself in the back of the head and get his arm down to the chest position, as in the Knox shooting.

I.D. Arnold, the Knox chauffeur, repeated in his testimony the same story he had told investigators, including Lillian's meeting with Collins. He confirmed he had driven Lillian to Beaumont, where Hiram was waiting in his Pierce-Arrow sedan. He said Knox had planned a hunting trip to Mission, Texas, but changed his mind. The next day he drove Mr. and Mrs. Knox to Houston, where they stayed in the Brazos Hotel. When he delivered luggage to the hotel room, he said he saw Lillian using a typewriter. Arnold said he drove the couple back to Hemphill for a conference with Internal Revenue Service agents.

Mrs. T.C. Carson, the wife of Lillian's secretary, was the first witness for Lillian. She said she had a conversation with Hiram shortly before his death, in which he complained bitterly about being "robbed right and left by everybody." She said he told her that since he had sold the East Mayfield sawmill and other properties in Sabine County, he felt despondent and "did not feel like living any longer."

J.M. White, a 77-year-old friend of Knox, said he had known the victim first at Livingston, then at Bronson, and finally at Hemphill. He said Knox sent for him shortly before his death "because he wanted company." He said Hiram told him he was despondent and "not at all himself." Later in their talk, White said Knox referred to his holdings in Mission and offered White a job in a gambling house he owned. When he was questioned by Hamilton, White admitted he had been in jail frequently, saying he had been charged with everything "from horse stealing on up."

The two letters found in Knox's pockets—addressed to

Knox's mother, Mary Knox, and his attorney, E.J. Mantooth of Lufkin—created considerable discussion and turmoil in the courtroom.

Alford said he turned the letters over to Sheriff Nations. When Nations was asked for the letters, he said they had been given to prosecutor Hamilton, which prompted King to demand that the letters be produced. Hamilton refused, contending the letters would wait until the following day, Friday. Justice Pratt agreed, touching off a round of outcries from Lillian's defense table.

"They can't afford to show these letters because they will exonerate the defendant," thundered King. "There are rumors abounding in this county that the woman is a murderess; that she shot and killed her husband. She is not afraid of this court or any other court. She wants to exonerate herself before the people who love her. We are not afraid of this court," he shouted.

King continued: "The court is not playing fair with us. We don't give a continental what sum of bond the court may fix. There are other courts we can appeal to and get justice, the fairness and right that every American is entitled to. I charge there is a concerted effort to suppress the evidence in this case."

Apparently unmoved by King's arguments, Judge Pratt overruled the objection and said it was up to the state to offer the letters. His decision produced another outburst, this time from Chester Collins. He declared: "If I were judge, I'd resign my office before I would suppress evidence favorable to the defendant." Pratt looked squarely at Collins and told him: "You will sit down, Mr. Collins, or I will hold you in contempt and send you to jail." Collins sat down.

When the argument was over, the defense rested without

putting Lillian on the stand to deny the charges. The prosecution also rested its case without introducing the two suicide notes.

The court ordered a recess and at 3:30 p.m. it was reconvened to give the defense and prosecution attorneys an opportunity for closing arguments. When they had finished, Pratt said: "I have carefully considered all of the testimony and exhibits produced during the two days of this hearing, and all the arguments of the counsel, and I hereby find that the State has produced ample evidence that Hiram Knox was probably murdered by the defendant on November 26, 1922. I am therefore handing the defendant, Lillian M. Knox, over to the Grand Jury of Sabine County. Bond is hereby set at $25,000."

King, still smarting from Pratt's ruling in the matter of the two letters, told the justice: "Mrs. Knox indignantly refuses to make this bond, even though she is eminently able to do so. She is innocent, and the amount of the bond is outrageously high." He also announced the defense would file a petition for *habeas corpus* the next day in District Court, seeking to free Lillian on a reduced bond, and that she be allowed to remain free on a temporary bond of $2,500 pending a hearing.

Pratt granted King's request and Lillian's bond was lowered to $2,500, but she had to remain under guard at her home. Sheriff Nations led her away, not to the county jail, but to her home.

On Friday, January 5, still dressed in black, Lillian and two of her attorneys, King and Minton, pushed their way through a crowd and entered the courtroom of District Judge V.H. Stark in San Augustine, less than thirty miles northwest of Hemphill. There, they filed a petition for writ of *habeas corpus*, by which they sought to remove Lillian's case from the Justice of the Peace court to the District Court. District Attorney F.P. Adams was present to oppose the move.

King, who would later become prominent as a U.S. attorney in the Beaumont federal court, told Judge Stark he would offer no witnesses and gave him a hastily-typed transcript of the two-day preliminary hearing in Hemphill's Justice of the Peace Court. Judge Stark set the hearing for that afternoon.

When the court reconvened, E.J. Mantooth of Lufkin, Hiram Knox's attorney and the addressee of one of the purported suicide notes, appeared in the courtroom and asked for a conference in Stark's private chambers. When they returned to open court, Adams announced the state would be satisfied with a $7,500 bond for Lillian. Her lawyers objected, saying there was no need for a bond at all. "The defendant is a pillar of the community. She is not going to leave this jurisdiction," said King.

The lawyers and Stark went back to his chambers for another conference. When they returned this time, a $5,000 bond had been agreed upon. Lillian and her legal team then went to the Court Clerk's office, posted the bond, and left with a gathering of her followers. The bond was signed by five Hemphill men, including Dr. Tom Arnold, a witness for the prosecution. The other four were E.P Padgett, a lawyer; J.P. Fuller, who owned a drug store; L.E. Pratt, a merchant; and John W. Minton, attorney for the Knox family. Another press report listed S.R. Williams as a surety.

Adams was left behind to talk to the press. "Much remains to be done. Beyond that, I have nothing to say," he told reporters.

Lillian's group then went to the Moss Hotel in San Augustine for dinner. In the lobby, she paused to declare to the reporters who had followed her that "I feel like a bird just out of a cage," even though she had not been tried for the murder.

Spotting an elderly Hemphill man in the crowd, she asked: "How's the folks?" To a Houston reporter, she said: "It's too bad all this publicity isn't doing any good. Wouldn't it be fine if we could make some money out of it for the children's home of the Houston Salesmanship Club?"

At dinner, Lillian's group found itself seated next to District Attorney Adams. According to a newspaper report, S.R. Williams of Hemphill joked to Lillian, "I didn't even read that bond I signed. It might have been my death warrant." Lillian responded: "If it is, I'll serve your sentence." Adams was not amused.

That night, Lillian was at home in Hemphill to await a hearing by the Sabine County grand jury scheduled for March, 1923.

Meanwhile, a flurry of stories in the *Houston Post* and its competitor, the *Houston Chronicle*, attracted considerable attention among people following the case. *Post* reporter John McClendon, in a copyrighted story, discovered in the San Augustine County Clerk's files documents Hiram Knox had executed to Lillian on November 21, 1922, and recorded on November 22, four days before his death. The documents had been notarized in Houston on November 21, apparently while Hiram and Lillian were in Houston. The documents transferred all of Hiram's property, real and personal, "to my beloved wife Lillian M. Knox" and specifically included all proceeds from the sale of the Knox sawmill to Temple Lumber Company, one-half of which he had already conveyed to her on October 8, 1921.

King called a press conference and announced: "This deed removes all financial motive for murder on the part of my client. She is innocent and we will prove it." The *Houston Chronicle* reported that District Attorney Adams did not attach any importance to the documents. He said they were known and had been widely discussed in East Texas for more than a month.

The *Chronicle* also said something that, in the end, would be prophetic of Lillian's fate. "The attractive widow, who has been so generous to her acquaintances, has many more friends than enemies in Sabine County, and her attorneys were perfectly satisfied at this time to let it (the deed) go at that."

But in the weeks that lapsed between Lillian's court appearances in Hemphill and San Augustine in January and the grand jury term in March, the media's interest in her case gradually lessened. The reporters left East Texas, Hemphill once again became a sleepy county seat town, and Lillian kept a low profile on the advice of her lawyers. She remained secluded in her home and stopped giving parties. The prevailing sentiment in Sabine County was that, if the grand jury indicted Lillian, she would be acquitted if the jury heard the same evidence introduced in the examining trial.

However, there were still a lot of questions being asked at the lumber mill, in the cafes at Hemphill, and in the town's church socials. Why would she have murdered her husband if he had already left her everything he owned? Was there a new, undisclosed will which excluded Lillian as a beneficiary of the Knox estate?

The Sabine County Grand Jury—a panel of twelve citizens summoned by the District Court to receive complaints and accusations in criminal cases, hear the state's evidence, and return bills of indictment if they felt a trial should be held—convened in March, but delayed Lillian's charges to April.

When the Grand Jury met to consider the charge, they heard only five witnesses, Frank Cook, Dick Williams, former Sheriff George Alford, W.L. Harrell and A.M. Jones. Only one of the five, Alford, had testified in the preliminary hearing in Justice of Peace Court. Lillian herself did not testify and was not required to because of her right against self-incrimination. Even her lawyers were not allowed to speak. By law, a Grand Jury's proceedings are secret and no report is made of its deliberations or the testimony of witnesses.

On April 18, 1923, the Grand Jury returned a report on its proceedings to Judge Stark in the Sabine County courthouse as Lillian and her lawyers stood by to post a new bail in the event she was indicted. The minutes of the Grand Jury said simply of Lillian's case; "No Bill." It meant that the twelve Sabine County residents had decided not to indict Lillian for Hiram's murder. Judge Stark dismissed the case.

Lillian was elated. She told one of the few reporters present: "I knew I would be vindicated. It couldn't be otherwise."

Those who were closest to the case said District Attorney Adams had little hope of an indictment. Judge Minton said everyone on the grand jury was either an admirer of Lillian or had been a beneficiary of her generosity.

Minton, in an August 5, 1958, interview with historian Robert S. Maxwell, said he felt Knox's death was "clearly a case of suicide." Minton also disclosed he was called to the Knox home on the morning of the death before Dr. Arnold and Sheriff Alford, and was actually the first to see the body, even before the household servants, Lillian and her sister, or Doctor Arnold and Sheriff Alford.

Minton's interview conflicted with testimony given by other witnesses in the 1923 hearing in Hemphill. He said Knox was found slumped sideways in his chair, not his bed, and that two letters hinting suicide were found in his pockets. The reasons the letters were not entered as evidence, he said, was that the signatures were considered to be reasonable matches with Hiram Knox's handwriting and, without handwriting specialists, it would have been difficult to prove otherwise. The suicide notes disappeared and presumably they were given to E.J. Mantooth and Mary Knox—the supposed recipients—after the grand jury made its decision.

If Lillian Knox had been convicted of Hiram's murder and sentenced to death under Texas law, she could have been hanged in the Sabine County Jail less than a mile from her home. When the jail was built in 1903, it was equipped with a hanging tower because executions were done in Texas counties in those days. The last hanging, ironically, was done in 1922—the year of Hiram's death. The executioner was Sheriff George Alford and the man sentenced to hang was Marshall Shell, who was convicted in the murder of his stepdaughter. Shell's last words were, "Crime doesn't pay and I am paying for what I've done."

After the grand jury set her free, Lillian returned to her home in North Hemphill and the company of her children and servants, but her life would never be the same. She would continue to be surrounded for the rest of her life by tragic deaths, police investigators, judges and lawyers—but for different reasons than those she faced in Sabine County in 1922 and 1923.

The Sabine County Courthouse in Hemphill, as it appeared during a hearing on Lillian Knox's murder charge in 1922. *Photo courtesy, Cecil McDaniel*

Sabine County Attorney T.B. Hamilton, who laid out the state's case against Lillian Knox in a 1923 hearing. *Photo courtesy, Cecil McDaniel.*

Lillian Knox, second from left, joined friends on an outing in Galveston before husband Hiram's death. *Photo courtesy, Cecil McDaniel.*

9

A FADING REPUTATION. As time passed, Lillian Knox faded from public view. She remained largely secluded for several years in her home at Hemphill, but the world around her was changing.

The residents of Hemphill saw her from time to time, but not as conspicuously. She knew she was no longer regarded with admiration and, in some quarters, she was still considered a murderess. Within the lumber industry, where she had achieved recognition as perhaps the only female manager in the business, most of her peers suspected she was guilty, but that she had cleverly concealed any evidence that could have connected her with Hiram's death.

The grand jury's decision to "no bill" Lillian on the murder charge filed against her did nothing to quell the rampant rumors. In fact, they were intensified throughout East Texas, leading to new and frequently fanciful charges.

A Sabine County man known for keeping whiskey hidden

in the woods near Hemphill claimed he was near his cache on the night before Hiram Knox's death when a car drove into the woods on a dirt road, stopped, and several voices were heard arguing. The man said he heard a shot and the car drove away. The story, however, was never reported to authorities.

Another belated story came from I.D. Arnold, Mrs. Knox's chauffeur in 1922. According to members of his family, Arnold moved from East Mayfield to Houston shortly after Hiram Knox's death. Before his death, Arnold told his family he had been instructed by Mrs. Knox on the evening before the murder to drive to Beaumont and bring her brother, Alex Marshall, to East Mayfield. He said he was also instructed to drive Marshall back to Beaumont early the next morning.

Ivy Buckley, a Sabine County resident in the 1920s and 1930s, told his family he heard "a lot of stories" that linked the deaths of several prominent Sabine County men with Hiram's Knox death. Buckley also said he was told that Lillian Knox gave a member of the grand jury five prized jersey bulls. One of the bulls had a temper, "but Mrs. Knox could walk up to him and tap him on the nose with a stick and he'd get out of the way," said Buckley. The grand juror's wife was afraid of the bull, Buckley said, so her son put on his mother's clothes and beat the bull, hoping it would make the bull afraid of the woman. "But, one day, she was killed by the bull," said Buckley.

Buckley, who died December 30, 1997, at the age of 86, said "there was an element in Sabine County who thought the sun rose and fell on Mrs. Knox; she had them all charmed, they never believed she did anything wrong. And she was a good-hearted woman, too. There was a story about a little teenage girl who was bitten by a spider. Mrs. Knox heard about it, and paid for all the nursing it took for the girl," said Buckley

Temple Lumber Company made improvements to the East Mayfield sawmill and some of the community's buildings, which it also owned, but the town no longer looked to the Knox family for guidance and assistance. Thomas Louis Latane Temple and his son Arthur Temple, Sr. became the town's new patriarch family.

On February 23, 1923, while Lillian Knox was battling the murder charge against her, the East Mayfield sawmill was badly damaged in a fire, but was rebuilt by the new owners. In February, 1932, a Beaumont newspaper noted that the big mill normally employed 350 men and cut 125,000 board feet a day "though it is operating at present two-thirds time."

In 1937 another fire destroyed most of the mill and lit up the night skies around Hemphill—a glow that could be seen from miles away. Arthur Temple, Jr., recalled standing on the porch of the old Knox Lumber Company office as a young man and taking a snapshot of the fire. This time, with America in a depression and lumber sales declining, the Temple family decided not to rebuild the mill. Most of East Mayfield's employees were given jobs at the Temple sawmill in Pineland. Some were transferred to another Temple mill at Diboll, south of Lufkin in Angelina County.

The two fires and others in neighboring communities played havoc with the lumber-based economy of Sabine County and the surrounding area. In addition to the East Mayfield fires in 1923 and 1937, a sawmill owned by Kirby Lumber Company at Bronson was destroyed in 1928 and another Kirby mill at Browndell, in neighboring Jasper County, burned in 1928. As a result, Sabine County's economy, which relied heavily on lumbering jobs, suffered greatly as the Great Depression descended on America.

When the Hemphill and Bronson sawmills went out of business, so did the Lufkin, Hemphill and Gulf Railroad, which depended on lumber shipments. The line had been operated in the late twenties and early thirties by the Texas South-Eastern Railroad of Diboll, also owned by the Temple interests, but the Bronson-to-Hemphill tracks were eventually abandoned when Temple Lumber Company exchanged logging trucks for the short line and its connecting tram lines. The railroad's tracks from Hemphill to Beanville, in the northeast edge of Hemphill, remained in place in the early 1950s and logs were banked along the tracks for loading and transportation to the Temple mill in Pineland.

Most of the East Mayfield homes which had been built by the Knoxes for its managers and employees were sold one by one by Temple Lumber Company to people in search of homes. Marguerite Eddings and her husband, a barber, purchased one of the homes on Barber Street. "When we came here from Alvin, Lillian Knox had already left town, but there was a lot of stories about her; most people felt she had helped a lot of people with her money," said Mrs. Eddings. Two other East Mayfield homes were purchased by Blanche Toole and her sister and moved closer to Hemphill. Their father, John Williams Fenley, was employed at the East Mayfield sawmill and as a logging foreman while the mill was operated by Temple in the early 1930s.

Lillian Knox's mansion in East Mayfield, once regarded as the largest and finest residence in Sabine County, became the property of the Temple family and was eventually demolished. Mrs. Toole said Lillian Knox's house in Hemphill, where she and her family resided after the Knox sale to Temple, was eventually purchased by Carol Arnold, a businessman who was later killed by his partner on the Sabine County courthouse square in the 1930s. The house was later destroyed by fire and a new home was built on the site by Mrs. Arnold

The city of East Mayfield, which had been incorporated on January 8, 1919, as a result of the Knoxes' interest, was disincorporated on November 20, 1949, when a petition was submitted to the Sabine County Commissioners Court. When the election was held, twenty-eight people voted to dissolve the town and twenty-three voted to retain the town's incorporated status. As a result, the town eventually became a part of the City of Hemphill.

Much of Lillian's property was in the hands of John W. Minton, her Hemphill attorney and a valued friend, but what portion was used to pay her legal fees and what part was held for her in a trust arrangement—possibly to shield her from future liabilities—was unknown.

In a detailed and lengthy deed executed on January 4, 1923, the day before she entered the courthouse for a preliminary hearing on her murder accusation, Lillian deeded to Minton lands in Sabine, San Augustine, Polk, Bowie, Harris, Hidalgo and Kleburg counties in Texas. The deeds covered nearly 12,800 acres—including nearly 740 acres originally owned by Sabine River ferryman James Gaines, a signer of the 1937 Texas Declaration of Independence. The deeds included most of the lands which had been conveyed to her by Hiram before his death. In addition, three other legal documents were executed by Lillian and Minton.

On March 15, before the grand jury hearing, Hiram's "catch-all" deed to her of November 21, 1922, was recorded in San Augustine County. The same deed was recorded in Sabine

County on March 20, 1922. In May of 1925, Hiram's will, witnessed by Minton, was filed by Minton in Sabine County. It had left everything to Lillian. On May 19, 1925, Hiram's deed was admitted to probate and Lillian was appointed executrix. On October 8, 1925, Lillian executed her oath of executrix and signed it in Hidalgo County, Texas.

No explanation was given why Lillian and Minton waited two years to file the will, but presumably they wanted to be sure no additional charges would be lodged against her and put the Knox estate at risk in the courts. Portions of the land, amounting to 8,917 acres were sold in later years to Temple Lumber Company as the company expanded its holdings in Sabine and adjoining counties, and some lands were sold to other parties.

In the 1930s, when the U.S. government created four national forests in East Texas, some of the old Knox lands became a part of the Sabine National Forest, which now covers lands in Sabine, Shelby and Newton counties. Some of the lands were replanted with new pines by the Civilian Conservation Corps, a Depression-era recovery agency, but most were allowed to reseed and regenerate themselves naturally. Today, the lands are now covered with some of the most scenic pine forests in East Texas. Other old Knox lands were flooded when a long stretch of the Sabine River was dammed in 1967 to create Toledo Bend Reservoir on the Texas-Louisiana border.

In 1924, Lillian traveled to Livingston, obtained a court order on July 25, and had the bodies of Colonel Knox, Grace Eva Knox (her husband's first wife), and her stepson, Willie Knox, removed from Forest Hill Cemetery, placed on a train, and shipped to Texarkana where they were reburied beside her husband's grave in a large family plot in Woodlawn Cemetery.

The Knox plot adjoined one containing the remains of her mother, Daisy E. Marshall, and two infant children. She then sold the Livingston cemetery lots for one dollar each. Lillian apparently intended to return to Texarkana to have monuments placed on three transplanted graves, but for some reason—perhaps one associated with her declining finances—she failed to do so.

One intriguing story told in 1987 by a former Knoxville resident living in Polk County said that when the three bodies were loaded onto the railroad car for the trip to Texarkana, John Frisby, the man who had been charged with Willie's murder, was the conductor of the train and was seen sitting atop Willie's coffin, waving to viewers as the train left Livingston.

In the years following the decision of the Sabine County grand jury to free her, Lillian also spent time in Beaumont, where the Knox family rented several homes, and in Mission, a community in Hidalgo County. Lillian reportedly invested in businesses owned by her brother, Alexander Marshall of Beaumont. In 1923, Marshall was the owner of a radio and automobile equipment business located in the lobby of the Wiess Building in Beaumont. In 1929, he was in the same location, but was involved in real estate, insurance and rentals. Marshall reportedly married a Spindletop oil heiress and continued to live in the Beaumont area. Lillian also reportedly invested in a Ford dealership at Mission.

In Beaumont, Lillian also talked about building a new sawmill in East Texas and returning to the lumber business. But East Texas, like the rest of America, was in a depression, lumber markets were just as depressed, and most of the region's large virgin forests had been cut.

Two of Lillian's sisters also lived in Beaumont. Clara Marshall married Spencer Blain, who became prosperous with

Blain's Cash Stores, a series of stores founded by his father, James S. Blain. Evelyn Marshall was married to Walter I. Troup, owner of a Beaumont drug store. The two families were apparently active and prosperous in the 1920s. Another sister, Fannie Marshall, married Jack Watson, a lumber dealer in Houston.

Lillian made occasional trips back to Hemphill to visit with her old friend and lawyer John W. Minton. "She always came under the cover of darkness, presumably because she did not want anyone to see her," said Minton's grandson, John H. Minton, Jr., an attorney in Tyler. "She was usually in a black car driven by a chauffeur," said Minton. As a young boy in the late 1930's, Minton often visited his grandfather's law office on the second floor of a bank building on the northeast corner of the courthouse square in Hemphill. There, he always found candy inside a shoebox kept in a filing cabinet in the office. "I was told Mrs. Knox left the candy, and kept it replenished for me," he said.

Ivy Buckley passed down to his children a similar story. "In the thirties, while I was working on a school building at Pineland for the Works Progress Administration, a friend and I were walking down Gravel Hill Road when her black limo, driven by Slick Waller, pulled up, and she offered us a ride. While we sat in the back seat, she talked to us and said she had been working on some oil leases. She also told us she was planning to run for State Representative from Sabine County," he said. However, she never made the political race.

In 1927, Lillian left East Texas and moved to Dallas with five of her six children, Hiram, 16; William, 15; Marcene, 10; Merton, 9; and Mary Marguerite, 7. The children were enrolled in North Dallas schools.

Even though she had little financial resources, she still had her old audacity and personal charm. She leased a two-story colonial mansion at 3717 Rawlings in an upscale neighborhood and started building a reputation as a *grande dame* of Dallas. She filled her leased mansion with fine mahogany furniture, damask draperies, imported china, and sterling silver. Some household items were remnants from her better days in East Mayfield; others were rented. She told a Dallas newspaper reporter that she was a member of a prominent and wealthy family and had received a good education. She claimed she and her sisters had attended school at "Miss Mason's School in Tarrytown, New York."

Lillian lived on a grand scale with black servants, lavish dinner parties, fine food and vintage wines. But there were few reciprocal invitations from those she invited and courted. Dallas wasn't Hemphill; the metropolitan city was full of wealthy families with good manners, bearing and status. Lillian, on the other hand, spoke with a rural accent, her grammar was poor and her tastes were considered gaudy. Dallas society shunned her, not only because her Hemphill past had been reported by the Dallas newspapers only five years earlier, but because she simply didn't fit into the town's tonier circles. Son Hiram said in a 1991 video interview that his mother "lived with a guy and they had an office across the street from the Adolphus Hotel, where they peddled fake stock bonds."

Lillian's reasons for moving to Dallas, when she could have lived anywhere, are not clear, but they likely involved Mary Knox, her mother-in-law and the grandmother of her children. Mary had remained in Dallas in 1912 when she and Colonel W.H. Knox decided to end their marriage. The divorce had left Mary with enough financial resources to live comfortably, if not lavishly, in Dallas. In 1924, she built a modest but pleasant house at 5746

Mercedes Street in Dallas and furnished it with quality furniture given her as a part of the divorce settlement.

By now, Mary was the only surviving member of the Knox family which had moved from Wisconsin to Dallas around 1900. In 22 years, she had seen the deaths of Grace Eva Briggs Knox, her daughter-in-law, in 1908 ; her first grandson, Willie Knox, in 1911; her former husband, Colonel William Hiram (Will) Knox, in 1913; and her only son, Hiram, in 1922. All but the Colonel had died unexpectedly under tragic circumstances. She had also lost her only daughter, also named Grace, to tuberculosis in 1900.

In 1931, an article about Lillian appeared with her photograph in a book entitled *The Rio Grande Valley and its Builders*. Books such as these were regularly assembled in the 1930s by vanity publishers who sold the articles and photos to people and businesses seeking to enhance their reputations and images. Such articles were always positive, laudatory and tended to stretch the truth.

The article said "Mrs. Lillian M. Knox, Dallas, Texas, resides at 3717 Rawlings Street in a stately mansion in which are found the true lines of the Southern Colonial home, and where her friends are always cordially welcomed."

The article did not mention Lillian's slain husband Hiram and implied that she was the wife of Colonel W.H. Knox. It said: "Mrs. Knox was among the pioneers of the Valley. The Knox Lumber Company opened the first large lumber yard, with a store in connection, at Mission. The late W.H. Knox, pioneer ... white pine lumberman and railroad builder, who came to Texas forty-one years ago, recognized the executive ability of Mrs. Knox and made her secretary and manager of the business and, although young, she proved herself most capable. That was the beginning of her business activities."

Lillian was further described as "one of the largest landowners in Texas, owns vast acres of timberland in this state and a large acreage in Mexico and is also interested in one of the largest independent oil companies in the United States."

The article continued to paint Lillian's past with laudable achievements. She was purported to be the only woman in the U.S. to build a standard charter railroad, the only woman in the U.S. to build and manage a three million dollar sawmill, one of the largest in the U.S. She supposedly employed 3,000 men in the sawmill for seven years and, according to the article, "her orders were carried out without question." The article said the *American Lumberman*, in a feature story June 8, 1918, paid "high tribute to Mrs. Knox as a patriotic, practical woman of the South, and she is the only woman ever given recognition" by the journal.

But the article did tell some truths. It mentioned Lillian's charitable work, her patriotic work during World War I and her work with children, woman suffrage causes, churches and the poor.

As Lillian's resources continued to decline for one reason or another, she once again showed her resourceful side. She converted her leased Dallas home on Rawlings Street into an upscale boarding house and housed her children in the attic. They were often told to enter the attic by climbing a large oak tree and walking across a plank "cat walk" to an attic window, presumably to avoid mingling with Lillian's boarders and diners.

There were soon enough paying customers to continue the lease and maintain Lillian's lifestyle. Her boarders complimented Lillian on the luxurious ambiance of the house. The dining room, with its homestyle cuisine, proved popular with lawyers, doctors, professionals and politicians who flocked to the house for informal noon luncheons as relief from Dallas' more formal restaurants and hotel dining rooms.

But strange people began to show up at the house at all hours of the day and night. Police were often called to deal with disturbances in the neighborhood. One such incident was a publicized police raid in search of a fugitive criminal. As a result, Lillian's respectable boarders began to move out and the diners went elsewhere for their meals.

Forced to supplement her income, Lillian found employment as a housekeeper in 1934-35 at the St. George Hotel, once ranked among Dallas' fanciest hotels. During the time she worked there, the St. George was known for its gambling games, operated by Jack Binion, who died on January 10, 1934, when a gull-winged Stinson monoplane crashed in White Rock Lake near Dallas. Only 21, Binion was already prominent in gambling and bootlegging circles, according to the *Dallas Morning News*. Jack was the brother of Benny Binion, whose life was chronicled in Jim Gatewood's biography, *Benny Binion: The Legend of the Gambler and Mob Boss.*

According to St. George hotel records, Lillian left late in 1935 and went to Baltimore when the St. George was purchased by B.F. Whitaker, the reported head of Dallas' gambling rackets in the 1930s. Whitaker had a portion of the St. George demolished, the rest remodeled, and renamed the building the Hotel Whitmore. Whitaker installed in the Whitmore a hidden ladder running the length of the building from its penthouse to the bottom floor, in case he needed a fast getaway.

Lillian apparently did not remain long in Baltimore because she was back in Dallas within a few months and was working in

1936 as a housekeeper at the Whitmore, which, according to police, had also become a notorious gangland and gamblers hangout. One news article said the hotel was also populated by "a running rampage of women." The Whitmore was demolished in the winter of 1974.

Mary Knox, nearing the age of 90, had for years blamed Lillian for the 1922 death of her only son Hiram. She had been cool to Lillian's arrival in Dallas in 1927 and did not welcome her into her home. But, perhaps because of her kind feelings for her grandchildren, her hostility began to thaw and Lillian was occasionally welcomed into Mary's home, about three miles from Lillian's home. Mary lived alone, shunned servants and preferred a simple lifestyle. Mary Knox had a special warmth for Lillian and Hiram's youngest child, Mary Marguerite, the namesake of her grandmother.

In the autumn of 1936 the older Mary Knox's attitude toward Lillian Knox changed abruptly when she told her neighbor, Mrs. Frank Rutherford, that she was planning to change her will. She said she planned to leave the bulk of her considerable estate to Reynolds Presbyterian Orphanage in Dallas. The only provision in her will for Lillian and her children would be five dollars each. Mrs. Rutherford had understood that Lillian and the children were to have been substantial beneficiaries in the old will, but were now being disinherited.

The reasons soon became clear. In October, 1936, Mary swore out warrants accusing Lillian of forging Mary's name on five checks totaling $1,155. When police went to arrest Lillian, she had disappeared. But, despite her apparent absence, Lillian was indicted on January 2, 1937, on five forgery counts.

On January 23, 1937, Mary swore out two more warrants

in Big Spring, Texas, against Lillian's son, William Hiram Ambrose Knox, 26, accusing him of forging her signature to two more checks. After graduating from high school, Hiram had worked in Dallas for several years, and then accepted a job with the Civilian Conservation Corps in Big Spring.

When police attempted to serve the warrants for Hiram's arrest, he, too, had vanished. During the weeks that followed, Lillian and her son remained in hiding while the other Knox children remained in the Dallas boarding house under the care of Lillian's two remaining servants, an elderly black couple.

The two fugitive Knoxes, however, were on the edge of bigger troubles with the Dallas police.

A painting of the Knox Lumber Company sawmill at East Mayfield. *Photo courtesy, Arthur Temple, Jr.*

Lillian Knox in a photograph taken early in 1919. *Photo courtesy, Merton Knox family.*

Young Willie Knox, whose grave was moved from Polk County to Texarkana in 1924, with a friend in this photograph made before his death in 1911. *Photo courtesy, Merton Knox family.*

A Texas Historical Marker commemorates the history of East Mayfield. *Photo, Bob Bowman.*

The offices of Temple Lumber Company in Pineland. The company purchased Knox Lumber Company and most of its assets in 1921 and 1922. *Photo courtesy, Cecil McDaniel.*

The mill pond of Knox Lumber Company at East Mayfield, as it appeared in 2002. *Photo, Bob Bowman.*

A Texas Historical Marker stands at the entrance of Hemphill City Cemetery, where many of Lillian Knox's Sabine County friends are buried. *Photo, Bob Bowman.*

A company-owned employee house at East Mayfield in the 1920s. *Photo courtesy, Cecil McDaniel.*

10

THREE LAWYERS. As Lillian Knox's life was becoming more complicated in Dallas during the 1930s, changes were also occurring in the lives of three key figures involved in the Hemphill murder scandal less than a decade earlier.

E.J. Mantooth, the Knox attorney who had been the recipient of one of Hiram Knox's suicide notes, was winding down his legal career in Lufkin. Noted for his taciturn demeanor, he never mentioned the 1922 suicide note Hiram Knox supposedly addressed to him, according to his children, and rarely discussed his relationship with the Knox family.

In the 1930s, during Lillian's residence in Dallas, Mantooth was resting on a career of achievements in Lufkin. He was one of Lufkin's founders in 1882, had served as an early mayor of the city, had organized the city's first telephone company with Dr. A.M. Denman, and was a respected attorney and businessman.

Mantooth had a long and interesting legal career and the

Knox death in East Mayfield wasn't his only brush with a murder mystery.

Late in the evening of March 3, 1913—nine years before he was thrust into the Knox case—a dynamite blast rocked downtown Lufkin, destroying a railroad depot owned by the Houston, East and West Texas Railroad. Glass was shattered in buildings blocks away and several boxcars burst into flames.

A Wells Fargo employee was crippled for life and 24-year-old Frank Parsons, an employee of the depot, was presumed dead after investigators discovered a burned body inside the wreckage. Parsons, reported the *Lufkin Daily News*, "was looked upon as a gentleman of the highest degree of excellence."

But as investigators kept looking, the explosion took on an air of mystery. The body turned out to be that of an animal.

Representing the insurance companies and the HE&WT, Mantooth suspected that Parsons had not died in the explosion. The depot employee's fate became the talk of Lufkin and East Texas for three years. Some people, like Mantooth, felt he was living elsewhere. But most people in Lufkin were convinced he had gone to his death faithfully on duty at the depot.

A high court agreed with the common theory and, after a legal battle, Parsons' stepmother collected a significant insurance settlement and Parsons was declared legally dead.

On January 4, 1916, a telegram arrived in Lufkin with this message: "Am leaving Los Angeles this afternoon for Lufkin. Will bring Frank Parsons with me. E.J. Mantooth."

The telegram hit the town like a thunderbolt. A movie theater flashed the message on its screen. The *Lufkin Daily News* reported the bombshell in large headlines and the news spread throughout Angelina County.

When Mantooth and Parsons arrived, more than a thousand people showed up to meet him at the railroad depot. As Parsons was led through the streets to jail, the crowd shouted a chorus: "That's him. That's him." The fire department's gong sounded the signal for District 36, the same signal which rang the night the depot exploded.

With the help of a private detective, Mantooth had found a letter written to the clerk of a Galveston court seeking information about the depot case. Parsons, it was disclosed, had sent the letter using an alias. Further evidence showed Parsons had enlisted in the U.S. Army and was given an honorable discharge.

When he was confronted by Mantooth in California, Parsons said: "Judge, I am glad it is all over."

But, like Lillian Knox's plight, it wasn't.

Parsons' trial for felony arson had to be held in Center, the county seat of Shelby County, on a change of venue when his attorney argued successfully that the accused could not get a fair trial in Lufkin.

While Mantooth's persistence had settled the mystery of the railroad agent's disappearance, the dynamiting of the depot was another matter. A Center jury wrote the final chapter to one of East Texas' most bizarre cases by finding Parsons not guilty.

He walked out of the courtroom and was never seen again. What really happened on the night of March 3, 1913, in Lufkin is still a mystery—just as was Hiram Knox's death on November 26, 1922.

Mantooth died on May 19, 1944, at the age of ninety-two and was buried in Lufkin's Hillcrest Cemetery beside his wife Caledonia, who had passed away August 10, 1928.

John William Minton, the Knox family lawyer who organized Lillian Knox's murder defense team, was still active as an attorney until the 1950's and corresponded frequently with the Knox widow, even as she moved from town to town and state to state.

From time to time, Minton also sent Lillian checks representing the sale of timberlands in East Texas and corresponded with her during the time she lived in Chicago. She also pleaded with him for money.

Born in 1875 at Geneva, Minton started his career teaching school and later working for a Bronson merchant. His son, John Henry Minton, also an attorney who in 1925 joined the senior Minton's law practice in Hemphill, was born in 1904.

While attempting to swing aboard a moving railroad coach, Minton fell and the train's wheels crushed his foot and severed the toes. A physician wanted to amputate the foot, but Minton refused. The foot healed, but Minton was unable to stand on his feet for any length of time, ending his teaching career.

In 1910, he ran for a seat representing Sabine County in the Texas House of Representatives and, while serving in Austin, he attended law school at the University of Texas, earning a law degree. Returning home in 1912, Minton and A.D. Hamilton, another Hemphill lawyer, opened a law practice associated with the father of former Texas Lieutenant Governor Ben Ramsey in San Augustine, although Minton and Hamilton practiced law in Hemphill.

At the time of his death, Minton and Hamilton owned some lands together, according to a grandson, John Henry Minton, Jr of Tyler. The elder Minton—often addressed as "Judge Minton"—died on April 19, 1959 and was buried in

Hemphill's City Cemetery.

Priscella Pratt Zeppa of Tyler, who lived in Hemphill as a young girl, remembers the elder Minton "as a distinguished, well-dressed, and soft-spoken man" who commanded wide respect in the community.

Minton's Hemphill home, which he built in 1922 before the Knox scandal, was sold to an East Texas businessman and moved to a knoll of land several miles east of Garrison in Nacogdoches County. Lillian Knox once claimed she had built the home for Minton, but probably meant that Minton's fees from Knox Lumber Company had enabled him to pay for the house. When he passed away on April 19, 1959, Minton owned only two tracts of land he acquired from the Knoxes. They were 17.5 aces where his house was located, acquired from Hiram in May of 1921, and an adjacent ten acres acquired from Lillian in 1926.

Judge Minton's son, John Henry, continued to receive correspondence from Lillian Knox. In most of her letters, Mrs. Knox had usually pleaded for money, starting as early as 1950 when she lived in Chicago. "Grandfather was always sympathetic to her, even though she was a conniver. My father wrote back to her that Granddad had died, and that would be the end of any money being sent to her. My dad seemed to take great pleasure in 1959 in writing her that the money from the Minton family had ended, and that she should forget it," said grandson John H. Minton, Jr.

In her correspondence, Lillian had assumed the name of Margaret Costello, but she apparently had warm feelings for the Minton family. "She often sent us very expensive gifts with the sales tags still attached, but we never knew if she had bought the gifts herself, or had gotten them from some of the wealthy people she took care of. We always received big boxes of gifts. On one occasion, as a teenager, I got a baseball glove and I remember on another occasion she sent my mother some heavy sweaters useful only in snow country," said Minton's grandson.

"On another occasion, she asked for $100 and wanted the money sent to a woman in Chicago. She had enclosed some sealing tape and asked that the tape be wrapped around the money inside a pre-paid envelope, presumably so it could not be opened without Lillian's knowledge. On another occasion, she asked that some money be sent to her in charge of a hardware store in Chicago," said Minton.

Minton's grandson said Judge Minton believed that Knox's death was a suicide, but concurred that the gossip around Hemphill had always persisted that Lillian was not only responsible for Hiram's death, but for his mother's 1937 murder in Dallas as well.

Chester Boone Collins of Lufkin, the Lufkin lawyer whom Lillian claimed was once her husband and who had represented her in the alleged murder of her husband, had become one of Texas' most famous criminal attorneys in the 1930s.

Collins, the son of a Lufkin dairy farmer, was practicing law in Lufkin in 1926 when Reverend J.R. Nutt, pastor of the First Baptist Church, persuaded him to represent Reverend J. Frank Norris, a Fort Worth fundamentalist Baptist preacher charged in the sensational shooting death of D.C. Chipps, a friend of Fort Worth Mayor H.C. Meacham, following a prolonged controversy involving Norris, Meacham and anti-Catholicism.

Norris was indicted for murder on July 19, 1926, and Collins became a member of his defense team. The sensational case also made him one of Texas' most sought-after criminal lawyers. Using some of the same aggressive methods he exhibited in Lillian Knox's hearing in Hemphill, Collins was able to win an acquittal for Norris on grounds of self-defense. Norris, who ministered to a congregation of an estimated 6,000 Baptists, often heaped praise on Collins during church services and called for "love offerings" to help pay Collins' legal fees. Norris was so grateful to Collins that he recommended his services to important individuals and organizations in Fort Worth. As a result, Collins moved to Fort Worth in 1927.

In the 1960s, Lillian attached some blame for the Knox family's financial downfall to Collins, when he represented Knox Lumber Company. She alleged he was not aggressive enough in protecting the company's interests in Washington as it was being pursued for apparent tax liabilities. "I knew Collins was the cause of this because I had quit him," she said in a letter to another attorney, Edward Barrett of Chicago. She said Collins "often talked of what the government could do to millionaire lumber and timberland owners, and it was a hard fight. My good husband could not endure the heartache of the government's treatment. I told him we could go to Washington and get it all cleared up, but Collins told him he would lose. Collins knew what banks we had money in..(he) did not go to Washington to handle the Knox case. He planned or thought he would get the other lumbermen's case." According to letters and conversations with Barrett, Lillian's animosity toward Collins increased as she grew older. She alleged Collins gained possession of "a fortune in fine diamonds...and we were partners in timberlands that I had furnished money to buy..." While she admitted Collins "had a fine legal brain," she also claimed "he tried to kill me twice."

Lillian may have had more than a business relationship

with Collins. Before his death, Eck Prud'homme, who worked for the Temple lumber interests in East Texas, often told his friends, including Arthur Temple, Jr., that he frequently functioned as Lillian's "messenger boy." In the 1920s, he said Collins' wife, Chloe Mantooth Collins, called him and asked him to deliver a message to Collins at the Knox home in East Mayfield. Prud'homme said he arrived at the Knox home, asked the servant for Chester, and was led upstairs to a large sleeping porch, where Collins was sprawled on a feather bed, dressed in silk pajamas.

Eddie McFarland, a friend of Collins, said Collins had a wide reputation as a "ladies man." He described Collins as "a very handsome and vain man who had tailors make his dress shirts from cloth imported from Ireland and wore $1,200 suits. "He had the ability to look good in anything, even old blue jeans," said McFarland, a Lufkin attorney. "Women were attracted to him, and I am sure Mrs. Knox was," he said.

Lillian's claim that she was married to Collins has not been verified by any documents, although the union could have been a common law marriage, which was frequent among couples in early East Texas. McFarland doubts the Knox attorney was ever married to Lillian, but he said their relationship led to friction among Collins and his wife. "Chester told me that Mrs. Knox, during her heyday, commissioned a Kansas City artist to paint a large, full-length portrait of him. Chester hung the portrait in the foyer of his Lufkin home and rigged above the painting a light so it would illuminate the painting when the front door was opened. "One day Chloe, who was an expert marksman, took a pistol and outlined Chester's body in the painting with a series of well-placed shots," said McFarland. Shortly after the incident, Collins divorced Chloe, but McFarland said he continued to love her.

"Several times a year, he would drive to Lufkin, pick me up and we'd drive out to Hillcrest Cemetery, where Chloe was buried after her death in 1958. Chester would sit there, mourn her, and talk about the good ol' days," said McFarland. "He loved squirrel dumplings and every Christmas, when he came to town, he'd persuade the chef at the Angelina Hotel to cook the dish for him," said McFarland.

McFarland said Collins often called Lillian Knox "the widder Knox" and said he was grateful to her for urging him to buy land as an investment. He also said Lillian was one of his most loyal clients.

Collins died in Fort Worth on October 20, 1960, at the age of seventy-two. He was buried beside Chloe in Hillcrest Cemetery.

Hemphill attorney John W. Minton, who assembled Lillian Knox's defense team in 1923. *Photo courtesy, Cecil McDaniel.*

E.J. Mantooth of Lufkin, an attorney for Hiram Knox. One of the two suicide notes supposedly written by the lumberman was addressed to Mantooth. *Photo courtesy, Mantooth family.*

11

MURDER IN DALLAS. On Saturday morning, March 13, 1937, the *Dallas Morning News* splashed a four-column headline across its front page with the news of another murder in the Knox family: "RICH WOMAN RECLUSE BRUTALLY KILLED BY INTRUDER; SAFETY BOX IS MISSING."

The victim was Mary Knox, 91, the mother-in-law of Lillian Knox, and the last of five members of the William Hiram (Will) Knox family which migrated to Texas from Wisconsin around 1900. All five were now dead.

The murder was discovered the day before when Mrs. Hugh Boedeker, who lived down the street from Mary's home at 5746 Mercedes Street, noticed that the elderly woman had not made her usual early-morning walk. It was as much a routine in her life as tending her rose garden and feeding her backyard chickens.

When she went to Mary's home Mrs. Boedeker observed

that the home's window shades were down from the night before. She was concerned, but assumed Mary was absent from the home.

Around 5 p.m., Mrs. Boedeker noticed that Mary Knox had not made an appearance and the shades were still down. She walked to Mary's house, noticed the morning newspaper on the porch, rang the bell and received no response. She tried to open one of the two front doors, but it was locked. She returned to her home and called Mrs. Knox on the telephone, but still had no response. She then called Mrs. Frank Rutherford, a close friend of Mrs. Knox, who lived about two miles away at 928 South Haskell, and explained her concerns.

When Mrs. Rutherford arrived with her nephew Charlie Steger, they rang Mary's bell, knocked on the door, and walked around the house until they found an unlocked window. Steger crawled in and opened the back door for Mrs. Rutherford and Mrs. Boedeker. As they walked through the house, they noticed untouched eggs and a pot of coffee on the kitchen table. Nothing seemed out of place in the kitchen.

When they reached Mary's bedroom, they found signs something was wrong. The drawer of a marble-topped dresser was open and most of its contents were spilled on the floor.

In the parlor, they found the body of Mary Knox, lying on her back in a pool of blood, her tongue blackened and protruding from strangulation. Her right arm was angled in a defensive position over her face. She had been brutally beaten in the head and a towel and a cotton apron were twisted tightly around her neck. Her hands were covered with blood.

Above her, on the parlor walls, were photos of her husband William Hiram (Will) Knox, her son Hiram and his first wife, Grace Eva, and her grandson Willie. Like Mary, all had died under tragic circumstances.

Shaking with revulsion, Mrs. Rutherford immediately called the police and then alerted the neighbors. By the time the police arrived, the house was filled with gawkers.

When Dallas police detective inspector Will Fritz and detectives Jack Archer and Buck Stroud arrived at the Knox home, they sealed off the crime scene and found that the skull of the elderly Knox widow had been crushed in four places by a blunt object which was suspected of being a flat-sided bar with sharp corners or a jagged pipe. She was wearing a cotton nightgown, a cotton dress, several petticoats, an old sweater and slippers.

Inspector Fritz and his team found that one of the front doors (the one Mrs. Boedeker had not tried to open) was unlocked, as well as the window through which Steger had entered the home. They theorized that Mary's assailant had knocked on the front door as she prepared her breakfast. When he was inside, the killer choked her to prevent her from crying out and then bludgeoned her to death.

Mary had apparently tried to escape from her murderer. A trail of splattered blood led from the front door to the living room. A long and narrow, but indistinct, bloody footprint of a man's shoe was found on the corner of the living room rug near the body. The footprint was cut out of the rug and preserved for evidence. Although the detectives didn't know it at the time, similar long and narrow footprints had been found November 26, 1922, on the ground outside Hiram Knox's home when he died of a single gunshot near Hemphill.

Stroud and fingerprint expert Harry Riddell found two smudged prints on the large dresser in Mary's bedroom, showing a "spatulate thumb and forefinger," both unusually large, and fingerprints were taken. A half-smoked, hand-rolled cigarette of brown paper was found in the fireplace. Since Mary was not a smoker, it was assumed the butt had been thrown there by her killer.

Since there were no signs of forced entry, Fritz was convinced Mary knew her attacker or attackers and had admitted them into her home. Mrs. Rutherford, who knew Mary better than most of her friends, said the slain woman would have never opened her door for a stranger and, in recent months, had been keeping all of her doors locked. It was a departure from a previous attitude, which was reflected in a statement she made earlier: "Why should I want to keep my doors locked. If a prowler should come in, I want the doors open so I can get out."

Mrs. Rutherford also told the police Mary had distrusted banks since the depression-era bank collapse. She said Mary always wore a money belt stuffed with cash and kept large sums of money in a locked steel box in her house. Fritz and his team searched for the money belt and money box, but found neither. They did find several bonds amounting to $11,000 hidden throughout the house, a cashier's check for $1,600, and six dollars in cash. Mrs. Rutherford felt (and Fritz concurred) that the killer or killers had known about the missing items, and had killed Mary to prevent them from being identified.

Fritz also theorized that the assailant had hidden on the front porch, which was screened by shrubbery, after the murder and robbery and had fled from the scene that morning when no one was watching. No other clues or evidence were found. Fritz was sure Mary knew her killer, but he wondered aloud why the killing was so brutal. Why the vicious beating? Why the strangulation?

Next, the inspector went outside to search for any evidence. He noticed that the front porch was screened from the street by shrubbery, affording concealment to anyone entering or leaving by the front door.

As the search of the Knox home continued, the telephone rang. When one of the detectives answered, he heard a man's gruff voice say: "Who is this?" The policeman answered, "This is a police officer in charge." The man hung up. Similar calls and hang-ups occurred while the house remained under guard. The calls were never traced.

While the investigation of the crime scene continued, Inspector Fritz summoned the coroner and in due course the body was removed to the county morgue to await the inquest. Finally, Fritz locked up the house, sealed off the entire premises under heavy guard, and left. But for several days Fritz returned to the neighborhood in the early morning hours to see if there were any strange persons or occurrences in the vicinity at that time of day, but found nothing unusual except for the unruly and noisy crowds ghoulishly milling around the Knox bungalow. He thought breakfast was an unusual time of day for a brutal murder in a quiet residential neighborhood.

While Mrs. Knox was described as a recluse, Mrs. Rutherford issued a statement on March 15 which refuted the description. She said the elderly woman loved to have friends and neighbors visit her home. "Neighborhood children used to sit on her porch and talk with her by the hour. There was one girl who frequently played the piano for her. She spent many hours with her flowers and tended a flock of chickens. Just two weeks before her death I went with her to buy some more chickens for her flock. She lived alone because she preferred to, but many people do that and are not recluses," said Mrs. Rutherford, who had lived with Mrs. Knox as a young girl. Mrs. Rutherford and a police guard took turns feeding Mrs. Knox's chickens.

Newspapers in Texas, particularly those in Dallas and Houston which had covered the murder of Hiram Knox fifteen years earlier, wrote about Mary Knox's murder for days, pushing the Spanish Civil War and Amelia Earhart's landing in Hawaii off the front pages. The *Dallas Morning News* called the slaying "one of the most baffling mystery murders in Dallas County in many years."

Reporters flocked to the Dallas police department to cover the story. One story in the *Dallas Times-Herald* disclosed that Dallas County Sheriff Smoot Schmid had been searching for Lillian Knox, 45, the slain woman's daughter-in-law, since January 2 on the five forgery warrants sworn out by Mary Knox in October of 1936. Lillian's son Hiram, 25, was also being sought by Big Spring authorities on two more forgery warrants sworn out by Mary on January 23 in Big Spring. All of the alleged forgeries were on checks drawn on the First National Bank of Dallas, bearing signatures purported to be those of Mrs. W.H. (Mary) Knox and endorsed to "Mrs. Lillian Russell."

Meanwhile, on a lead supplied by a reporter, Inspector Fritz interviewed Eleanor Goebbels, who owned a Dallas employment agency. She told him Mary Knox had called her on the Monday before her death, asking the agency to provide a suitable companion to live with her. Knowing that Mrs. Knox had for years preferred to live alone, Mrs. Goebbels said it would take time to locate a suitable person. Uncharacteristically, Mrs. Knox told her: "I'll take anyone, just as long as she is clean and a Christian. I'm an old woman, alone and afraid. I need someone."

One of the first friends of the slain woman to be called was Dr. Floyd Poe, pastor of City Temple Presbyterian Church, of which she was a member. Mrs. Knox had given significant sums of money to the church. She had also donated a lot at Bryan and Carroll streets as a site for a church, but the property was later sold as a site for a bakery and the proceeds given for religious purposes. The *Dallas Times Herald* also reported that "Mrs. Knox is thought to have had some thirty rent properties as her only source of income."

Mary Knox was described by her family as "a real person with no hint of phoniness." During the height of the Great Depression, a relative's jersey cow died and lamented the loss to a relative. Mary learned of the loss and sent a $400 check to the family to buy another cow.

Reports of the gruesome murder in the quiet residential neighborhood of East Dallas caused hundreds of persons to congregate in the street near the Knox home at night. Many of them remained for days, hoping for a glimpse of the interior of the house. Children on skates climbed on the concrete porch and attempted to peek inside the windows. Drunks attempted to impersonate officers or newspaper reporters in the hope of getting inside. Automobiles drove slowly by the house, causing traffic problems. Such difficulties led police to place a heavy guard around the house and at one time fifty officers were assigned to keep the crowds from getting too close.

As the police continued their investigation of Mary Knox's death and Texas newspapers reported each new disclosure, the rumor mills were grinding away unmercifully in Hemphill and East Texas. Lillian Knox was not only being blamed anew for Hiram Knox's death in 1922, but was now being implicated in his mother's death in Dallas.

Inspector Fritz, meanwhile, was beginning to focus his investigation on Lillian and her son. But neither could be found.

Fritz put a 24-hour watch on Lillian Knox's colonial mansion on Rawlings Street, and on Sunday, March 14, a policeman saw the missing Knox son climb an oak tree beside the house and enter its third-floor attic by a board catwalk. Armed with the two forgery warrants from Big Spring, the policeman entered the house and found Hiram in his attic bedroom. He was arrested without resistance, photographed, fingerprinted and jailed. Two right shoes were taken from Hiram's closet and the house was searched for evidence from the murder scene.

Meanwhile, his mother had called Deputy Sheriff Bill Decker and Inspector Fritz by telephone to negotiate her surrender on the forgery charges. Inspector Fritz, who was unhappy with certain aspects of Lillian's surrender, told her to give up.

When Lillian heard of Hiram's arrest on the radio, she went to the Dallas County Sheriff's Office and surrendered to Decker on the forgery warrants. Decker, the press said, was "an old friend of the family." Like her son, Lillian was photographed, fingerprinted and jailed, but told reporters: "I want them to catch the man who did it," referring to Mrs. Knox's slaying.

On the following day, Monday, March 15, the *Dallas Times Herald* disclosed in a story that Mary had cut Lillian and her children out of a handwritten will found in her home. The will was dated January 28, 1936, twenty days after the five forgery indictments were lodged against Lillian. The will had been witnessed by Mrs. Rutherford and Adlin Ann Johnson and filed by attorney Charles McCombs in Dallas County Probate Court.

In the will, Mary bequeathed to "my son's wife, Lillian five dollars." Her grandchildren—Hiram, William, Marcene, twins Merton and Mary Lyllian, and Mary Marguerite—were also left with only five dollars each. Hiram said his grandmother "dropped us like a hot potato on account of one person," an apparent reference to his mother. He said his grandmother and mother "didn't get along."

The grandmother's will did not mention that the daughter Mary Lyllian was then living in a Catholic convent in Shreveport, Louisiana, where she had been left by her mother at the age of eighteen months.

Mary also gave Mrs. Sydia Rutter, a grand niece living in Piedmont, Kansas, two houses and lots on Catherine Street in Dallas and the remainder of her estate to Reynolds Presbyterian Orphanage House and City Temple Presbyterian Church in Dallas. Earlier in her life, Mary had also helped endow a tuberculosis hospital in New Mexico, where her daughter Grace Erma Knox had died some thirty years earlier. Mary's home on Mercedes Street was given to her sister, Mary Rosel.

The *Dallas Morning News* reported that Mary Knox's funeral services were held at City Temple Presbyterian church on Monday, March 15. She was buried in Grove Hill Cemetery, where she had earlier purchased a lot. Lillian and Hiram did not ask for police permission to attend the funeral. Other members of Lillian's family attended the funeral and it was reported that young Mary Marguerite Knox, who had reportedly developed a warm relationship with her grandmother, was particularly affected. In the middle of the police investigation, Mary "left for Beaumont where she will visit with relatives for at least a short time," reported the *Times-Herald*.

The *Dallas Morning News'* front page story included two

glowering full-length photos of Lillian and her son, plus another article reporting that Lillian's cook and chauffeur had also been held in custody for several days for questioning concerning Lillian's whereabouts. Police reported that the cook had told acquaintances that she was afraid there would be bloodshed in the Knox family soon. The police also questioned Lillian's chauffeur, Booker T. Sloan, but did not hold him. A man identified as a former roomer and friend of Mary Knox was also briefly questioned.

Meanwhile, Lillian and Hiram's fingerprints and Hiram's two right shoes, along with prints found at the murder scene, were sent to the Texas Department of Public Safety laboratory in Austin for comparisons.

On Tuesday, March 16, Lillian's troubles were intensified. While she sat in her jail cell, Sheriff Schmid served her with another warrant accusing her of passing a forged $178 check at a Dallas furniture store on November 7, 1936. She now had six forgery charges facing her. Prosecutors filed the new charge to hold Lillian in jail during the ongoing murder investigation because they doubted the forgery charges filed by Mary Knox could be used to detain Lillian since Mary was not alive to pursue the cases.

On the same day, the Dallas newspapers reported that Dallas homicide detectives had spent the previous day questioning Hiram before transporting him to Big Spring, 200 miles away, to face the two forgery warrants there, although authorities doubted Hiram could be held much longer either. Hiram was apparently the last person to see Mary Knox alive. He told Fritz he called on his grandmother at her home the day before her death, asking her to lend him $500. He said she told him to come back to see her "in a day or two." Hiram denied he was guilty of the murder. "I wish I knew who did it," he said, "I had no reason to kill my grandmother. I needed money, but there was no money in grandmother's house, only some bonds and stuff."

Hiram later told his family his grandmother told him she wanted nothing to do with Lillian and her children. He quoted her as saying: "Go away, you are her child. I don't recognize you."

On Wednesday, March 17, Inspector Fritz had an opportunity to question Lillian Knox in the county jail. He asked her about her activities on the day of the murder, as well as the night before. Fritz said Lillian was calm and composed, claiming she had been in Texarkana on the day of the murder. Fritz said she responded to his questions with plausible answers and near the end of the questioning he asked her who she thought had killed Mary Knox. Lillian replied: "An old enemy of the family did it, and I'm afraid for my children." Fritz said she seemed to express genuine fear.

The next day, when Fritz resumed his interrogation, he asked Lillian to identify the family enemy. "Oh, that's just a lot of foolishness," she said. "I just said the first thing that came to mind. Mary was a stingy old woman. She used to have hundreds of thieves and tramps do chores for her because she was too stingy to keep a regular servant. It looks to me as if she did that once too often."

Lillian refused to say more and Fritz ended the questioning. There was no request from either side for a lie detector test probably because the polygraph was not in general use in Texas at the time.

Fritz had also sought to question a Fort Worth attorney who represented Lillian in her business affairs when she lived at

Hemphill, but the attorney failed to appear. While the lawyer was not identified by newspaper reporters, he was presumably Chester Collins, who helped defend Lillian when she was accused of killing her husband Hiram in Hemphill. Collins was living in Fort Worth in 1937 and had become one of Texas' leading criminal lawyers. Collins did not represent Lillian in the Dallas murder investigation, but sent a member of his firm to talk to Lillian.

On Thursday, March 18, the *Texarkana Gazette*, which had been following the Knox murder because of Lillian Knox's past residence in the border city, reported that the Dallas police had received from four of Mary Knox's grandchildren, William, Marcene, Merton and Mary Marguerite, "undisclosed information voluntarily submitted." The story said the four grandchildren voluntarily called at the police headquarters

The murder was now consuming much of the police department's time. At least six detectives were assigned to the case and more than 300 telephone calls had been answered from the public.

One call came from a 33-year-old man who was taking a mail-order correspondence in detective techniques and was sure he could solve the Knox case. The man, however, was arrested when he deluged Inspector Fritz with news clippings about the crime after receiving his third lesson in the mail. The mail-order detective was finally released and dismissed as a crackpot when he confided that he was on the verge of solving the Lindbergh baby kidnapping case, which had been closed two years earlier.

By late March, Fritz had only one avenue left to explore. He was waiting for a state laboratory report on the shoe print and fingerprints found at the murder scene. In 1937, although there was no statute covering the subject, fingerprints were admissible in the courts, but with a requirement of eight points of comparison for admissibility.

On Saturday, March 20, Fritz received his report, but the contents were of little help. The thumb and fingerprints were too smudged to make a reliable comparison with prints made from Hiram and Lillian's fingers. The bloody shoe print from the carpet was too indistinct to make a comparison with Hiram's shoes.

Fritz had no hard evidence that Hiram or Lillian were involved. He had nothing to carry to a grand jury, much less a trial jury. He hadn't even found the missing cash or securities. And his two prime suspects had said nothing that would help him.

While authorities had no evidence to charge Lillian and her son with murder, they remained in jail on the forgery charges filed by Mary Knox before her death and the sixth charge lodged March 16. They also faced another legal problem. On March 1, Frank Josey, Jr., the owner of the mansion Lillian had been renting, filed a lawsuit seeking to evict the Knox family for failing to pay the monthly rent of $65 a month for December, January and February. On March 7, a Dallas justice of the peace, acting on Josey's lawsuit, approved a writ of restitution. When it was served on Lillian's son, William, at the Rawlings Street address, he retained attorney John Coffman to represent Lillian in the eviction and to help release her and Hiram from the Dallas jail.

Coffman persuaded the JP to delay the eviction so a settlement could be negotiated and on March 16 filed a petition in the Dallas County District Court for a *habeas corpus* writ that would require Sheriff Schmid to show cause why he was holding Lillian and Hiram. Responding to the filing, Judge Henry King set a writ hearing for Hiram on March 20, but declined to set one for Lillian until she had made bond in the forgery cases. District

Attorney Andrew Patton asked for a bond of $5,000 for each of the five forgery cases, but King set the bonds at $2,000 each. The *Dallas Times Herald* wrote on March 17: "Judge King made it plain that he would not agree to the release of either Mrs. Knox or her son until police had plenty of time to complete their investigation."

Coffman tried to help Lillian's case by suggesting, on her behalf, that the estate offer a $1,000 reward for the capture and conviction of Mrs. Knox's slayer. Police officials revealed they had received an anonymous tip from Alabama, informing them that Lillian's long-dead husband had correctly predicted his death, as well as his mother's, well before he was killed in Hemphill in 1922.

Lillian was soon confronted with additional legal problems. A Longview, Texas, clothing firm sued her for $174.35 for clothing bought on December 18, 1935; Opal Burford of Oklahoma City won a judgment against Lillian for allegedly taking a suitcase and clothing when she moved out of a boarding house in 1936; and deputy constables began removing from Lillian's leased mansion furniture she had purchased from two Dallas companies for $6,243, which she did not pay. Telephone and water services were also cut off at the house for non-payment and Lillian's boarders started moving out. While his mother remained in jail, young William Knox stood by helplessly, as deputies hauled away some of the last items of furniture left over from the family's days of prosperity in East Mayfield. A deputy constable described the furniture as "wonderful stuff" of the finest quality.

Lillian finally broke her silence and agreed to an interview with reporters at the county jail. She offered a $500 reward for her mother-in-law's killer, claiming she "had absolutely nothing to do with the murder." She added: "I have always been close to the family and even the estrangement that followed my husband's death never changed my feeling for them."

Lillian claimed she still owned land in fourteen East Texas counties. "Oil has been discovered on some of it, but it won't begin paying for about sixty or ninety days. If I could get out of here (the jail), I could raise some money," she said.

On Tuesday, March 23, after being held for nine days for questioning in the Knox slaying, Lillian was released from custody after her attorneys, John Coffman of Dallas and Otis Rogers of Fort Worth, forced a showdown in the courts. A day earlier, they made bonds totaling $12,000 in six forgery cases in which she was accused of forging Mary Knox's name to checks totaling $1,155 and Rogers had persuaded the Texas Court of Criminal Appeals in Austin for a writ of *habeas corpus*, alleging that Mrs. Knox was being held illegally in Dallas. When the showdown came in district court, Assistant District Attorney Gene Bailey advised Judge King that Inspector Fritz was finished with his questioning of Mrs. Knox and did not wish to have her held in custody any longer.

Lillian left the courtroom, saying she intended to set in motion plans to obtain the release of her son Hiram, who was still in jail at Big Spring. With Coffman's help, Hiram was released on bond and the forgery charges against him and his mother were dismissed the following month on Coffman's motion because the complaining witness, Mary Knox, wasn't available to complain.

When another suspect, a black man, was questioned and released by Fritz and other detectives, the investigation hit a dead end. Lillian and her family were evicted from their mansion and most of their furniture was attached and sold to pay their debts.

More than a month after Mary Knox's death, five of her grandchildren—Hiram, William, Marcene, Merton and Mary Marguerite—tried but failed to contest Mary's will, which had left them only five dollars each. There is no record that Lillian and her children ever collected their five dollar bequests. While newspapers in Dallas said Mary Knox "was reputed to be a millionaire," her estate was valued at only $75,000 by her estate's attorney, Charles McCombs.

By early April, 1937, Mary Knox's murder had been dropped from the front page of the Dallas newspapers. The *Dallas Morning News* reported on April 3 that police investigators "have tacitly admitted they are without a clue in the murder." Adopting a sympathetic attitude toward Lillian, the newspaper said she "became the unfortunate victim of circumstances after the aged woman's body was found and the word went out that the police were looking for her. She (Lillian) answered by giving up voluntarily and remained in jail while trying to make bond long enough for the police to find out everything there was to know about her...at the same time, she was not permitted communication with anyone who could manage her business affairs and two furniture companies seized more than $6,000 worth of furniture in her Southern mansion she operated as a boarding house."

On April 11, Inspector Fritz admitted to the *News* that, while the Mary Knox murder case was still open, he had no fresh leads. "Why, for all I know, the murderer might be the common variety or alley-running prowler," he said.

Lillian's son Hiram, meanwhile, was indicted on eleven new forgery charges by a Howard County grand jury in Big Spring. He was also charged with mail fraud in connection with alleged shortages in the funds of Templeton & Cannon, a post office building contractor which had employed him. While Hiram served a year in prison, he claimed he was eventually released and cleared of the crime when he convinced authorities his mother had framed him. He said he accepted the prison sentence because he couldn't bear the thought of sending his mother to prison. Hiram said he eventually entered Texas A&M University on a scholarship, joined the A&M Cadet Corps, became an engineer, married Mary Elizabeth Burns of Big Spring, and traveled throughout the world with Brown & Root Construction Company of Houston.

On April 23, 1937, the *News* reported that Lillian Knox made a trip to Livingston in Polk County, where she had entered the Knox family in 1910 by marrying Hiram. Her reason for traveling to the community was to sell 89 acres of land in Polk County to B. Frankurt.

Gradually, Lillian made other trips from Dallas to East Texas—most of them to Sabine County, where she had once been the county's reigning *grande dame* and its beloved Lady Bountiful. But, as she did, she became increasingly secretive, traveled mostly at night, and avoided using her own name.

Mary Axford Knox, divorced wife of Colonel William H. (Will) Knox. *Photo courtesy, Mrs. Frank D. Hollar Family.*

Mary Knox, the victim of a brutal murder in her Dallas home. *Photo, Master Detective Magazine.*

Dallas detectives examine a chest in murder victim Mary Knox's home. *Photo, Master Detective Magazine.*

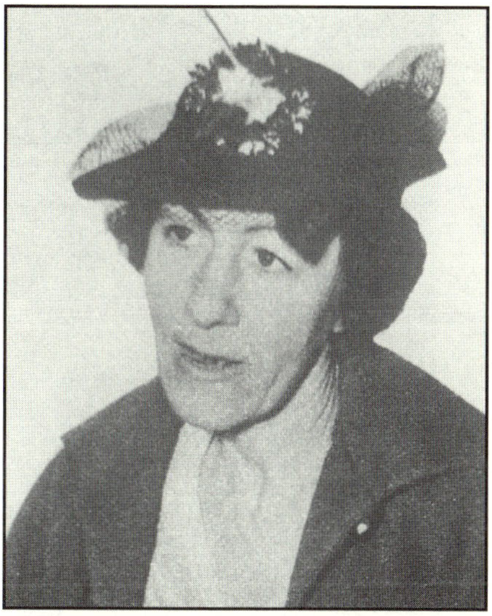

Mrs. Frank Rutherford, a close friend, discovered Mary Knox's body in her home. *Photo, Master Detective Magazine.*

Lillian Knox's photo appeared in a 1938 issue of *Master Detective* Magazine, which described the deaths of her husband and mother-in-law. *Photo, Master Detective Magazine.*

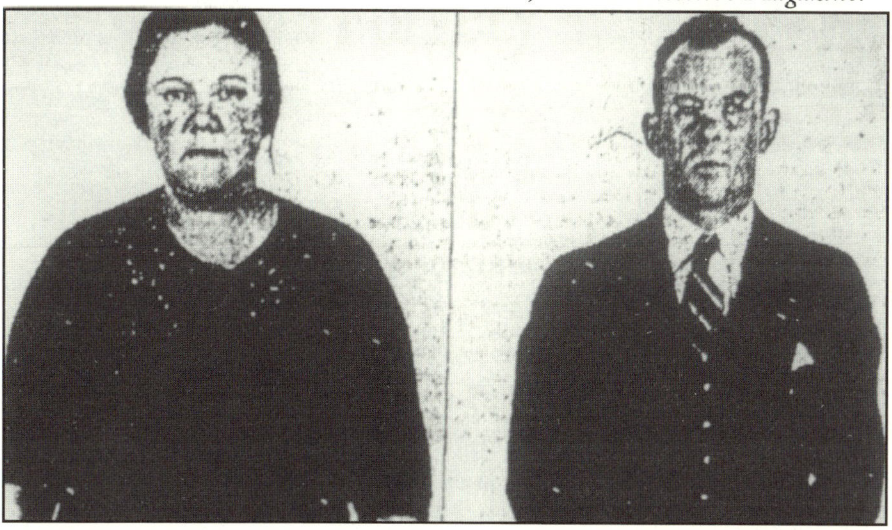

Lillian Knox and her son, Hiram William Ambrose, in police photographs made in Dallas in 1937. *Photos Master Detective Magazine.*

Workmen remove furnishings from Lillian Knox's colonial mansion in Dallas in 1937. *Photo, Master Detective Magazine.*

12

A PRISON SENTENCE. Following the death of Mary Knox in Dallas, Lillian Knox and her children went separate ways.

An irony of sorts also occurred in 1938 when Lillian's first husband, Roy J. LaRue, died in Texarkana at the age of 56. LaRue, who was married to Lillian for less than two years between 1906 to 1908, was buried in a Woodlawn Cemetery plot next to the one holding the remains of her second husband, Hiram Knox. LaRue's second wife, Effie, was buried beside him in 1941 at the age of 66.

Lillian apparently remained a close friend of LaRue and his wife, even to the point of having Hiram Knox's funeral in the LaRue home in 1922. LaRue, a marble worker at a Texarkana monument works, may have created a large marble monument and plot enclosure for the Woodlawn grave of Daisy Marshall, Lillian's mother, sometime during the Knox years of prosperity in Sabine County.

Around 1938, her reputation further tarnished by another Knox death, Lillian apparently moved from Dallas to Shreveport, perhaps, in a moment of remorse or guilt, to be present for an important event in the life of her youngest daughter, Mary Lyllian Knox, whom she had abandoned as an eighteen-month-old infant with the Daughters of the Cross Catholic Convent in Shreveport.

On August 15, 1938, twenty-year-old Mary Lyllian decided to embark on the process to become a Catholic nun. When she pronounced her perpetual vows on August 16, 1944, she assumed the name of Sister Mary Clement Knox.

Mary Lyllian had been born October 6, 1918, in East Mayfield as a twin to brother Merton (Bud) Knox, but their mother apparently separated the twins as small children, sending Merton to live with a family in Wisconsin. Lillian told him the family was that of an uncle, but he later found there was no relationship. Some of Lillian's children felt she had tried to sell him to the family.

While Merton was serving with the Air Force in the Pacific, the mother superior of the Catholic convent, nearing death in Shreveport, disclosed to Sister Mary that she had a twin brother. Sister Mary located Merton through military channels and wrote a letter to him while he was serving with the U.S. Air Force. "He was absolutely surprised," said Dorothy, Merton's wife. Sister Mary became a beloved part of the Knox children's families and visited frequently with her siblings until her death in 1988. "She was the sweetest and most fun-loving nun I had ever met," recalls Cecilia Fry of Auburn, California, a niece of Hiram Knox, Merton's brother. "She enjoyed laughing and was a wonderful person to be around, especially when it came to children," she said.

In the late 1930s, Lillian Knox also traveled to Big Spring, Texas, to see son Hiram and his wife, Elizabeth. Virginia Spears, Elizabeth's sister, described Lillian as a "very beautiful woman, with big brown eyes, well-groomed hair, and wearing lots of fox furs."

She also went to Sabine County, where she and her Hemphill attorney, John W. Minton, sold 215 acres of cutover timberland to Wendell Nethery of Milam for six dollars an acre. Nethery said she came to Hemphill in a chauffeured automobile and stayed overnight with A.A. Wilson of Geneva, who had worked at the old Knox sawmill in East Mayfield. While there, she asked Wilson to lend her seven or eight dollars. Wilson wrote her a check, but when it cleared the bank it had been raised to seventy or eighty dollars. Nethery said she also passed off a hot check to a Hemphill bank.

"She was a fairly large woman at the time, but she looked well off. I asked her how much money she got when the Knox sawmill was sold to the Temple family, and she told me eight and a half million. I asked her what she did with all that money. She laughed and said she had a lot of friends," recalled Nethery

The land sale was the second such transaction between the Netherys and Knox. Several years earlier, Lillian and Minton sold about 280 acres to Nethery's brothers, C.O. and C.A. Jr., for ten dollars an acre.

Another Sabine County resident, Cecil McDaniel said he served in the military with Lillian's son, Merton, at Fort Leonard Wood, Missouri, and met Mrs. Knox in the late 1930s when she came to Hemphill on business, again riding in a chauffeured automobile. "While she was in town, she bought some groceries from my boss, grocer Earsel Coussons, who had managed the old

Knox commissary store at East Mayfield. She paid for them with a check and it bounced, but Earsel shrugged it off," said Daniel.

Despite her entanglement in two murder investigations in fifteen years, Lillian Knox—now in her fading glory as Texas' Lady Bountiful—had served less than ten days in jail.

In 1922, when her husband Hiram was killed in his own bed, Lillian spent one night in the Sabine County jail when Sheriff George Alford charged her with murder. She was released the next day on bond and was eventually no-billed by a county grand jury.

In 1937, when Hiram's mother Mary Knox was found bludgeoned to death in her Dallas home, Lillian was kept languishing in jail nine days on forgery charges while police investigators explored her involvement in the murder. She was eventually freed without being charged in the crime. She also escaped the forgery charges.

But in 1939, Lillian's luck ran out.

On February 27, a Shreveport federal grand jury for the U.S. District Court for Western Louisiana returned a fifteen-page indictment against Lillian on eleven counts of using the mails to defraud. Using the alias of Helen Maulin, Lillian had written and cashed eleven phony checks amounting to $1,724.92.

The victims were three individuals and four banks in Louisiana and Texas. Seven of the checks were drawn on a nonexistent bank and all of them were written on bogus accounts between October 4, 1938, to December 21, 1938. The largest check was for $497.80; the smallest was for five dollars. Ironically, one of the checks was written on November 26, 1938, the sixteenth anniversary of her husband Hiram's death in Texas.

Lillian was arrested and, unable to make bond, dispatched to jail to await her trial. The court appointed G.M. Bedenheimer,

Jr., a Shreveport lawyer, as her attorney, indicating that Lillian—who once had a worth of millions of dollars—was now indigent.

Arthur Temple, Jr., of Diboll, the heir to the Temple lumber empire in East Texas, told a similar story about Lillian's forged checks. "Sometime in the late thirties, she owed our company a sum of money, say $5,000. One day, she came to my father with a large check, much larger than $5,000, and said she wanted to pay him back. She asked him to cash the check, take the money she owed the company, and give her the difference. The check bounced, and my father filed charges on her, and she went to prison," said Temple.

"My father, like others in the family, was intimidated by her," he said. He recalled a story told by Henry Temple, who ran the family's lumbering operations in Diboll, Texas. "In his home at Diboll, Henry had a bedroom overlooking the yard beside his house. On several mornings, Henry would wake up early, look outside, and see Lillian walking around his yard, apparently waiting for him to rise so she could ask him for money. It really unnerved Henry," said Temple.

In late September of 1939, as Lillian remained confined to her jail cell in Shreveport, she was stunned by the news that her youngest child, nineteen-year-old Mary Marguerite Signoria Knox, had committed suicide while living in Beaumont. Mary, who had been deeply affected by her grandmother's murder in Dallas, reportedly took her life September 20 during the despondency of breaking up with a boyfriend.

The suicide may have been the reason Lillian told Judge Gaston L. Porter that she was ill when she entered a plea of guilty during her arraignment on the forgery charges November 17. In a unusually subdued appearance, she withdrew her plea of not guilty and said she did not desire to be represented by an attorney. The sympathetic judge ordered her to remain in jail for treatment by a government physician until she could be sentenced.

Although Lillian had dismissed attorney Bedenheimer, he was in court on February 22, 1940, when Porter found her guilty of the charges and sentenced her to a five-year term in the U.S. Reformatory for Women at Alderson, West Virginia, on the indictment's first count, which involved a $350 bad check. The charges on the remaining ten counts were suspended, but Lillian was required to remain on probation three years after serving her five-year prison term.

On March 8, 1940, U.S. Marshal Louis E. LeBlanc delivered Lillian to the federal prison in West Virginia. She was logged in as Inmate No. 4744. She listed her age as only 42 and said she was born in 1897 (she was actually born in 1891). She had also gained considerable weight and now tipped the scales at 187 and a half pounds and stood five feet and three and one-fourth inches. She listed her occupation as a housewife. Her criminal history was blank; no mention was made of the two murder investigations or the previous forgery charges.

While in prison, Lillian wrote frequently to her old Texas associates from her cell. One letter directed to Henry Temple of Diboll asked him to send her an assortment of Christmas gifts—such as lipstick, stockings and cosmetics—for her fellow prisoners. Fearful Lillian would confront him when she was released from prison, Temple complied. On another occasion, she mailed a package of linens, apparently made in prison, to one of her children.

Lillian's sentence was scheduled to expire on February 21, 1945. Her term in prison was uneventful and she had only one "good time forfeiture" for mailing three uncensored letters to her sons in an envelope sent to the address of "Helen Raymond" in Camp Hall, Colorado. The Raymond woman had died and the envelope was returned to the prison. Lillian was penalized on December 7, 1943, with a forfeiture of ninety days good time, but on January 30, 1944—more than a year before she completed the full five years of her sentence—Lillian was released from the West Virginia prison. Seven and a half months later, Lillian's abandoned daughter, Mary Lyllian Knox, pronounced her perpetual vows as a Catholic nun and assumed the name of Sister Mary Clement.

By now, all of Lillian's children were adults and had moved to different parts of the country. Lillian was on the move, too, this time to Illinois, where she would again run afoul of the law.

Twins Merton and Mary Lyllian before they were separated at the age of eighteen months. *Photo courtesy, Merton Knox family.*

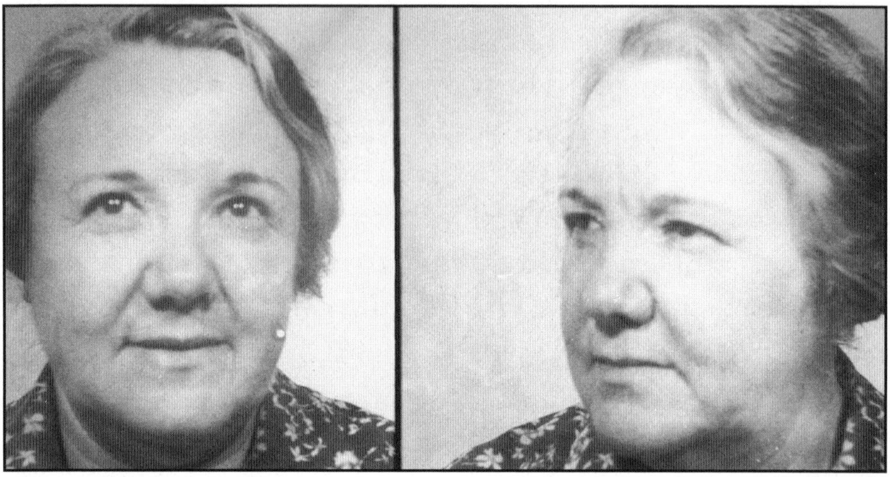

Lillian Knox's prison photographs when she entered a West Virginia women's reformatory in 1939. *Photo credit, U.S. Reformatory for Women, West Virginia.*

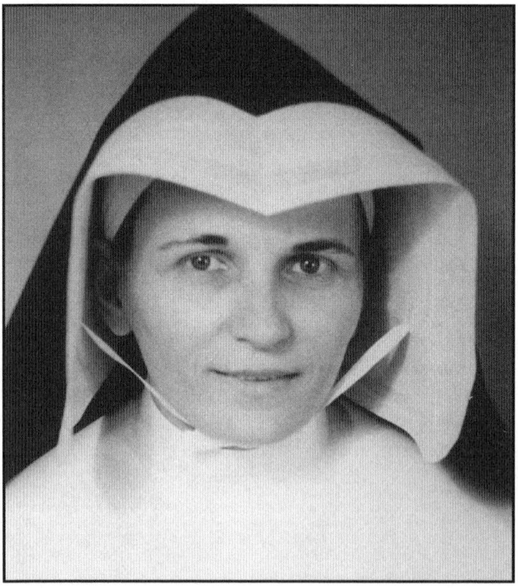

Mary Lyllian Knox, who was given to a Catholic convent in Shreveport by her mother, Lillian Knox, became a nun when she was 20. *Photo courtesy, Merton Knox family.*

13

LILLIAN IN CHICAGO. In March of 1957, as World War II veteran and attorney Edward J. Barrett sat in barber Pat O'Connor's chair inside Chicago's 100 North La Salle Building, O'Connor asked him if had ever handled any criminal cases.

As a 36-year-old lawyer practicing in Chicago's famed Loop with Rhea Campbell, another veteran, Barrett had been a lawyer in practice seven years. He and Campbell were eager to take on almost any case.

As O'Connor snipped at his hair, Barrett said he had handled some criminal cases, including one in which he successfully defended a young professional baseball player accused of rape, but he wasn't paid when the athlete skipped town without paying the legal fees.

O'Connor had once cut the hair of VIPs on the Twentieth Century Limited train, which streaked from Grand Central Station in New York to Chicago in sixteen hours between 1902 and 1967. The luxury rail liner had maids and *maitres d'hotel* to care for its celebrity riders such as movie stars, business magnates, and *bon vivants*. It was the grandest of America's trains and pampered its passengers as if they were Vanderbilts, who just happened to own the New York Central Railroad, which operated the Twenieth Century Limited.

O'Connor continued to snip away, saying: "A friend of mine, an older lady, has been in Cook County jail for five months on phony larceny charges. She really needs a lawyer to handle her trial."

"Can she afford a lawyer?" asked Barrett.

"Well, she hasn't got any money now, but I am sure she'll get some if you take her case. They're trying to force her to take a public defender, but she doesn't trust any of them," said O'Connor.

"Pat, why are you so sure she's innocent?" asked Barrett.

"I'm sure that if she had anything of Mrs. Rose Moir's, the woman she is accused of stealing from, it was given to her before Mrs. Moir died. I saw them together a lot and I saw how they treated each other. They were good friends. Margaret didn't have to steal from her."

O'Connor explained that his jailed friend was known as Margaret Costello, but was being prosecuted under her real name, Lillian Knox. O'Connor said he felt Lillian was using an alias in Chicago because she had been in trouble in her home state of Texas. O'Connor said he had known Margaret since 1952 when she moved into the Castle Hotel on Chicago's North Side. Lola McElroy, a friend of O'Connor, was the owner of the hotel, as well as a public stenographer, and had introduced the barber and Margaret. The trio soon became good friends.

McElroy, continued O'Connor, was a long-time friend of

Mrs. Rose K. Moir, the wealthy widow of Harry C. Moir, the founder and owner of the Morrison Hotel in Chicago's Loop. In its past the hotel had been one of the city's most prestigious hotels and the one-time headquarters of the powerful Democratic Party of Chicago. During its latter years, it became a hangout for the Chicago mob.

O'Connor said Mrs. Moir was an invalid, confined to a wheelchair, living at the Morrison Hotel, and in need of around-the-clock help. McElroy had introduced Margaret to Mrs. Moir, and the three of them often sat in the mezzanine of the Morrison, sharing memories.

With O'Connor's haircut finished, Barrett agreed to talk to Margaret. He made arrangements and drove to the old Cook County Jail, where Chicago's Criminal Courts were located. The story goes that Chicago's late Mayor Tony Cermak fobbed the property off on Cook County in the 1920s because he couldn't get rid of it any other way.

Having tried two cases in the out-of-the-way Criminal Court Building, including a visit to the men's section of the County Jail, Barrett remembered the stench and noise in the jail. He wasn't looking forward to his visit with Inmate Number 9504004, Margaret Costello, in the women's section. He discovered that she was strangely listed as Costello on the jail records, but had been indicted and jailed as Lillian Knox.

Barrett was frisked as he passed through the jail's system of checks—one for lay people and another for lawyers visiting clients—and proceeded by elevator to one of the floors in the women's wing. Briefcase in hand, he emerged from the elevator into a corridor nine feet wide. Beyond was a wall of iron bars, separating him and the women guards from the prisoners.

About one hundred women were milling around noisily in a day room, most dressed in prison uniforms consisting of light blue short-sleeved dresses. The jail house stench was overpowering.

As Barrett approached the bars, a heavyset older woman in her prison uniform separated herself from the crowd and came forward. The other prisoners quickly stood aside deferentially. "I knew that this old hen ruled the roost," said Barrett.

Standing five feet four and weighing at least 160 pounds, the woman had straight short hair, faded orange henna hair, going to white. Her facial features were round and coarse with a button nose and small brown gimlet eyes. "She was a little old, stumpy lady with coarse features and spoke with a country-accented twang. She looked like a Chicago bag lady, but she had an oddly commanding manner with a penetrating look," said Barrett

Margaret/Lillian looked Barrett over appraisingly and said: "I'm Margaret Costello. You must be Mr. Barrett."

Barrett confirmed he was and asked her where they could talk privately. Margaret/Lillian snapped her fingers at one of the women guards, who directed Barrett to a lawyer's booth adjacent to the day room. He found a chair, a telephone, and a plate glass window between him and a similar booth on the other side. As he sat down and readied his yellow legal pad and pen, Margaret/Lillian was ushered into the opposite booth.

Each picked up a telephone. Margaret/Lillian told him she had been arrested in New York City in October of 1956 on an Illinois warrant for grand larceny of $53,000 in bearer bonds belonging to Rose Moir at the time of Mrs. Moir's death in December of 1954 and for perjury in denying possession of the bonds in Probate Court while under oath. She said she had been assigned a public defender because she had no funds, but she wanted a private lawyer. Lillian had languished in jail for five months.

Beginning his questioning of Margaret/Lillian, Barrett heard that the bonds had been in a tightly-wrapped Christmas gift from Mrs. Moir shortly before her death, but Margaret/Lillian claimed she didn't know about the gift until she opened the package in New York long after the Probate Court and inheritance hearings. She said, therefore, she had not stolen anything and had not lied under oath.

Unable to communicate clearly with Margaret/Lillian in the hubbub of the women's jail, Barrett asked Margaret/Lillian to write down her version of the facts and send it to him. He agreed to take the case on a promise from Margaret/Lillian that she would have his initial fee of about $1,000 mailed to him by a lawyer friend. In a few days, he received a cashier's check for the fee from John W. Minton of Hemphill, Texas, the one-time attorney for Knox Lumber Company.

Soon, Margaret/Lillian began sending Barrett frequent and voluminous letters and notes, some typed, some handwritten. Some came by mail. Some were hand delivered by jail guards, who seemed anxious to do her bidding. Much of the rambling content of her writings was irrelevant to the issue of her case, but Barrett digested every word for fear of overlooking helpful information. He also picked up additional information from her two friends, Pat O'Connor and Lola McElroy, and obtained the Moir estate's version of the case from Moir attorneys Herman A. Fischer and J. Milton Guy.

When Barrett sat down to review what he knew about Margaret Costello and Lillian Knox, he discovered he had as a client a woman with a past unlike any client he had ever represented.

Like others in Texas, Barrett would find that Lillian Knox, despite her age, was still capable of captivating people. Her presence would continue to drop in and out of his life for the next nine years.

Chicago attorney Edward J. Barrett represented Lillian Knox on charges of grand larceny and perjury.

14

MARGARET AND ROSE. As Lillian Knox unfolded her story to Edward Barrett, he learned she had arrived in Chicago in 1944 after serving a federal prison term for forgery and larceny.

Leaving the U.S. Reformatory for Woman in West Virginia, she dropped her real name of Lillian Marshall Knox and assumed the alias of Margaret Costello. She lived uneventfully by finding occasional jobs as a practical nurse caring for elderly people, a profession which began her ascent as a millionairess in the Texas lumber industry nearly forty years earlier.

Now in her fifties, Margaret was stocky and strong enough to meet the physical requirements of lifting patients. Between jobs she lived in a succession of inexpensive residential hotels, mostly on Chicago's north side.

In 1952, while between jobs, Margaret moved into the Castle Hotel on North Kenmore Avenue, owned by Lola McElroy, a respectable older woman who was both a public stenographer and a hotelier. The two women soon became good friends, although they had different pasts and personalities. Lola was ladylike, well-spoken, slim, well-dressed and unassertive. Margaret was the opposite in every way.

While living at the Castle, Lola introduced Margaret to two friends, Rose K. Moir and Pat O'Connor, who had owned the barber shop at the old Morrison Hotel. Margaret was soon accompanying Lola and Pat on visits to Mrs. Moir at the Morrison. Mrs. Moir, who had been a widow since 1933 when her husband died, was in poor health, but remained mentally sharp.

In July of 1952, Lola telephoned Margaret with the news Rose had been allegedly kidnapped by her long-time attorney, Herman A. Fischer, and placed in the North Shore Mental Hospital in Winnetka, a northern suburb of Chicago, presumably because he, as her attorney, was trying to protect her assets, even though he could have, according to Barrett, opened a conservatorship in Probate Court. Rose had called Lola, pleading to be rescued from the institution. Lola asked Margaret to help because she was busy at the time.

The next day, Margaret showed up at Fischer's office and, using the persuasive charms honed from her Texas days as a businesswoman and socialite, persuaded the lawyer to arrange for Rose's release. Margaret and Pat then drove to Winnetka, retrieved Rose, and went to the Harrison Hotel, a residential hotel on Chicago's South Clark Street, where Margaret and Rose registered and stayed for about ten days. Margaret attended to Rose's needs, endearing herself to the widow. It is unlikely Rose knew of Margaret's checkered past in Texas.

When they left the Harrison Hotel, Margaret and Rose

moved to the New Lawrence Hotel, another residential hotel, on the north side of Chicago, where Margaret continued to care for the demanding widow. After two weeks of around-the-clock care, an exhausted Margaret returned to Lola's Castle Hotel while Mrs. Moir remained at the Lawrence with a new practical nurse.

In four or five months, the new nurse also left Mrs. Moir. Margaret and Lola then placed Mrs. Moir in a nursing home, the Waverly Manor, on North Sheridan Road, where they felt she would receive professional care. She continued to be visited by her trio of friends, Margaret, Lola and Pat.

During this period in early 1953, Margaret received a check in the amount of $11,021 made out to "Margaret Costello," from her Hemphill, Texas, attorney, John W. Minton. The check carried the legend, "Hemphill property sale," an indication that Lillian Knox was still receiving money from the Texas estate she had inherited as a result of her husband Hiram's shooting death in 1922.

In September of 1953, Mrs. Moir decided she wanted to leave Waverly Manor and persuaded Margaret to move with her into another residential hotel, the Bryn Mawr Apartment Hotel on North Kenmore Street, where they would remain until Rose's death on December 16, 1954.

Margaret charged Rose eighty dollars per week to stay with her, provided it was paid in cash. Rose later raised her pay to one hundred and fifty dollars, perhaps because she feared she might lose her live-in companion. Margaret told Barrett she earned her pay because Rose was not only incapacitated, but imperious, incontinent and demanding.

During the 15-month period, Rose and Margaret were visited by a number of people, including Lola, Pat and several physicians. Herman Fischer, Rose's attorney, also came by for a visit, but Rose refused to see him. Another frequent visitor was Harry D. Moir, Rose's adopted middle-aged son who, according to Margaret, lived on Chicago's Skid Row and stopped by Rose's apartment for occasional handouts from his mother.

As a wealthy woman, Mrs. Moir had a large checking account and a safe deposit box at Chicago's First National Bank, filled with securities. Rose wrote checks for her expenses.

"As a lawyer, I have been on both sides of estate controversies, representing both the estate and the so-called culprit and vice versa, and I learned from the police that the practical nurse/elderly person con game is the most prevalent white collar crime in a big city, largely because an elderly person is usually incapacitated, isolated and dependent on the nurse or caregiver," said Barrett.

Before the Knox case, Barrett had represented the estate of Wilmer J. Montgomery in a lawsuit against several Chicago banks, whom he accused of negligence in allowing the Montgomery family's maid to cash checks on the Montgomery account without asking for sufficient identification. Barrett's victory in the case resulted in a landmark Illinois law which required banks and savings and loan associations to seek more identification from individuals cashing checks on accounts other than those maintained by the individual.

Barrett's description seemed to be what was happening at the Bryn Mawr Hotel in 1953 and 1954, although it was never proven. Fischer claimed that Margaret, whom he discovered was Lillian Knox, had apparently persuaded Rose to sign blank checks to several pre-selected stores in the neighborhood of the Bryn Mawr. Margaret would then purchase items needed by the Moir household, plus large amounts of canned goods. Apparently in league with the owners of the stores, she filled in the checks for amounts much larger than the purchase, pocketing the difference in cash and presumably splitting the windfall with the owners.

The cartons of canned goods were delivered to Mrs. Moir's apartment, where they were stacked to the ceiling. Lola and Pat said they saw the stacks, but claimed they never suspected anything amiss. The strange part of the arrangement was that Margaret, with the help of two apparently innocent neighborhood men, Irwin Hayden and Nat Saletko, periodically shipped huge boxes of canned goods to indigent families in her old hometown of Hemphill, Texas. When Barrett later asked Margaret about the shipments, she shrugged it off with, "Oh, they all went to charity."

In a letter written to Barrett on August 15, 1957, Lillian claimed that Mrs. Moir not only knew of the accumulations of canned goods, but helped her repack them for shipment to "very poor families." She wrote: "Mrs. Moir sat in a large chair with a very heavy card table...we would take the groceries out of the boxes and place them in front of her, and she would...tell Ethel (the Moir maid) and I what to put in the boxes. The Express man came in and picked up the boxes right in front of Mrs. Moir."

That was the good side of Margaret; she always had a soft spot in her heart for poor people, especially those from her old sawmill days, when life was hard and jobs were demanding and dangerous.

The bad side, claimed Fischer, was that the canned goods scam and forgeries using Rose's checks, netted Margaret around a quarter of million dollars. "The odd part was if Margaret got all that money, no one, even I, ever found out where it went. She never lived high during her Chicago days," said Barrett.

As Barrett plunged deeper into the charges against Margaret, he learned that in January of 1954 Rose Moir asked Margaret and Lola to go to her safe deposit box at First National Bank Safe Deposit Company in Chicago, remove all of her securities, and deliver them to her. Armed with her authorization, they got a ride from Hayden, obtained entry to the box, and removed only the bonds, mostly bearer bonds which did not have the owner's name. When they delivered the bonds to Rose, she was angry because they had not obtained her stock certificates and ordered them back to the box to find them. They complied and Rose endorsed all of her stock certificates in blank, saying she had torn up her will and was going to establish a trust for her son Harry.

For the next eleven months, Margaret and Rose lived together and were visited from time to time by Lola, Pat and other friends. Rose, however, never called her attorneys and neglected to follow through with her plans to set up a trust for Harry. According to Margaret and Lola, Rose's stock certificates lay scattered throughout her apartment, along with hundreds of cartons of food stacked to the ceiling in every room.

In early November of 1954, about a month before her death, Rose managed to talk to Lola outside Margaret's presence. "Lola," she said, "Margaret has done so much for me. I have given her enough to live on the rest of her life."

Margaret also wrote a note to Barrett, confirming what Rose told Lola: "Mrs. Moir gave me two gifts in early November, 1953. The first was for a check for $35,000 or $40,000 for my birthday on November 15. I cashed $20,000 of that, paid $3,000 to my attorney, David Fisher, sent $10,000 to my son in Brazil (which he never got). I kept $10,000 in Louisville (Kentucky) bonds."

Later, Rose purportedly gave Margaret a small box and told her to take care of it, but asked her to wait until Christmas to open the gift. When Margaret offered to find some wrapping paper for the gift, Rose declined. She then showed Margaret a robe in the box and wrapped it in brown paper. Folded inside the robe, supposedly unknown to Margaret, was $53,000 in bearer bonds.

On December 16, Rose died and Margaret called her attorney, Herman A Fischer, to report her death. Fischer came to the Bryn Mawr apartment and, for the first time, saw the cartons of food. He also found Rose's torn-up will and told Margaret that Mrs. Moir had been crazy. He also ordered Margaret out of the apartment, but she did not go empty handed. Margaret told Barrett she later carried $200,000 in Mrs. Moir's endorsed stocks to the funeral home and handed them to Fischer.

Margaret said she bought and paid for Mrs. Moir's burial clothing and sent $500 each to Rose's brother and niece so they could travel to Chicago for the funeral. They kept the money, according to Margaret. On the day of the funeral, Margaret said only she, the undertaker and the minister attended Rose's last rites. Even Harry, her son and sole heir, was absent.

15

MORE LEGAL TROUBLE. A few weeks after Rose Moir was buried in Chicago, Lillian Knox—still using the alias of Margaret Costello—found herself immersed in a new wrangle with the courts. This time, her antagonist was Herman Fischer, the lawyer for the Moir Estate.

Appearing before the Probate Court of Cook County, Illinois, Fischer asked for and received a court order appointing him the "administrator to collect," which enabled him to search for Rose's missing bonds and any other assets which had gone astray. He was, in essence, an officer of the court, and went after Margaret with zeal and retribution.

In early January, 1955, Fischer petitioned the Probate Court for a "citation to discover assets" against Margaret, alleging that $53,000 in bearer bonds were missing from Mrs. Moir's safety deposit box at the First National Bank. Margaret was served with the citation and summoned to court to testify under oath before presiding judge Robert Jerome Dunne. She was represented by attorney David Fisher (no relation to the Moir estate lawyer). Asked if she had any assets belonging to Mrs. Moir, Margaret said "no." According to Margaret, she knew that nobody, including Judge Dunne, believed her.

Margaret then did what she had done on several occasions in Texas. She disappeared without telling her lawyer, violating a court order to remain available for further testimony. She claimed she left Chicago because her doctor told her to "take a rest."

Fischer was infuriated. He quickly obtained a court order freezing all of Margaret's assets, most of which were held under assumed names. They included a Merrill Lynch investment account in the name of Mary E. Costello, another Merrill Lynch account in the name of Daisy E. Marshall (her mother's name), a Mercantile National Bank account, also in the name of Daisy Marshall, and the contents of a storage compartment owned by Reebie Storage Company.

Despite Fischer's best efforts, Margaret/Lillian/Mary/Daisy had vanished without a trace. On June 23, 1955, Judge Dunne found her in contempt of court for disappearing while under a citation order. He also found her guilty of misuse and appropriation of estate funds. On the same date, Fischer obtained a warrant for Margaret's arrest for grand larceny allegedly committed on December 15, 1954, and for perjury in the probate court.

Fischer later replicated copies of Rose's will, which she had torn up, alleging that the originals had been stolen from her safe deposit box. The probate court admitted the will and found that Harry C. Moir, Jr., was Rose's sole heir. The court also issued an injunction prohibiting the transfer of missing Borg Warner and Sears stocks and appointed Fischer and his son Carlton coexecutors of the estate, as specified in Rose's will.

Armed with his legal tools, Fischer went to work. He eventually collected $450,000 in stocks and bonds, plus $80,000 in U.S. Government notes, for the estate. He said he was unable to find 402 shares of the Borg Warner stock and 500 shares of the Sears stock, but had obtained replacement certificates. On March 28, 1956, Fischer filed his final account as the "administrator to collect" and received $20,000 in fees for himself and $20,000 for J. Milton Guy, a fellow attorney. Lillian claimed she knew Guy from her Texas days, and was fearful he was familiar with her past, but Guy denied it.

Meanwhile, the probate proceedings were suspended because of the disappearance of the star witness, Margaret Costello. For 18 months, she continued to elude Chicago's police.

Using a new alias, Hazel Bentley, Margaret made her way to New York City and managed to stay underground from April of 1955 to October 30, 1956, staying most of the time in Room 626 at the Woodstock Hotel with a young Greek sailor who had jumped ship in New York. During her flight from the law, she cashed a few coupons from the bearer bonds, claiming that while in New York she finally opened Mrs. Moir's Christmas gift box containing the robe and found the $53,000 in bearer bonds.

Hazel's sailor boyfriend was apparently unaware of the windfall because Hazel kept the bonds to herself, put them in a suitcase, and stored them in a public locker at Grand Central Station. The bonds remained intact, but Hazel occasionally cashed a few of the attached coupons . She felt no guilt because the bonds were a gift from Mrs. Moir, and Mrs. Moir had told her so.

Eventually, however, the coupons' paper trail led to Hazel's hotel room at the Woodstock. Armed with Illinois warrants for her arrest, New York Police officials arrested her at the hotel October 30, 1956, breaking up her love nest with the sailor. For the first time since her release from federal prison in West Virginia, Lillian Knox reverted to her real name. Indignantly, she told the arresting officers the bonds were a gift and were rightfully her property. She then led the police to the public locker and produced the bonds.

On November 2, 1956, Chicago police officer Helen McGill arrived in New York City to return Lillian to Chicago on the arrest warrants. Lillian had no attorney and had waived extradition. McGill went directly to the police lockup where Lillian was held, processed the warrants, and picked up the bonds in the evidence room, signing for them and placing them in her briefcase. She then went to the women's detention room and met Lillian. A brief conversation ensued in the presence of two New York police officers, in which Lillian admitted nothing and did not see or touch the bonds, which remained locked in McGill's briefcase.

Exhibiting a docile and ingratiating attitude, Lillian left the jail handcuffed to McGill. They flew back to Chicago, where Lillian was placed in the custody of the Cook County Sheriff's Office. McGill then delivered the bonds to the State Attorney's Office, where they were held as evidence against Lillian.

Lillian, meanwhile, was placed in the women's division of the Cook County Jail at the corner of 26th Street and California Avenue in Chicago. She had no funds for bail because all of her accounts had been frozen by Fischer, so she remained in jail. On December 28, she was indicted for perjury and grand larceny. She had no attorney and David Fisher had withdrawn as her attorney in Probate Court.

In March of 1957, she wrote Judge Wilmer Crowley, chief judge of the Criminal Division: "I have four grown sons, three of them fought in World War II and the Korean War. My daughter, Sister Celestine, is a nun in Livorno, Italy (Sister Mary Clement Knox was actually a nun in Louisiana). I have five grandchildren. Mr. Fischer attached my bank account, including $12,000 of my own money and my clothes. I need an experienced lawyer. My sons are trying to raise money for me."

Lillian also told Crowley: "I had voted for you each election year (in which) you ran. I think it is a compliment for you to have juries, for their seeing you will give you more solid votes. However, you will never need additional voters. You will hold office as long as you desire it."

Lillian never heard from Crowley and no one came up with the money for a lawyer. As a result, the Cook County Public Defender was appointed as her attorney, which is a routine procedure when a defendant has no funds.

At this point, prodded by barber Pat O'Connor, Chicago attorney Edward Barrett entered Lillian's life. He would prove to be her salvation in much the same way John W. Minton did in 1923 when he assembled a legal team to extract her from charges that she had murdered her husband in Sabine County, Texas.

Barrett knew that Illinois law required the state to bring a defendant to trial within 90 days of the indictment or arrest or the defendant goes free—unless the defendant causes the delay. He found out Lillian had caused not one, but several delays. She was assigned for trial to four different judges, but rejected all of them. She told Barrett she had heard through the jail grapevine that each one of the judges was considered a "hanging judge." Her protests against the legal system had kept her in jail for almost five months.

As he began to work on Lillian's case, Barrett poured over six pages of tightly-written yellow, legal-sized sheets of paper written in her handwriting, retelling the story of Mrs. Moir's secret Christmas gift of the bearer bonds wrapped in a robe. "It didn't really matter whether I believed her or not; this was her story, and her only defense, as I saw it at the time. There was no one to contradict her. And Lola McElroy had added some credence to Lillian's story by testifying at the inheritance tax hearing that Mrs. Moir had said she had given Margaret (or Lillian) enough to live on for the rest of her life," said Barrett.

As it turned out, however, Barrett did not have to use the defense. He found another way to free Lillian and, at the same time, turn the tables on the lawyers who had pursued her.

16

IN COURT AGAIN. As Edward Barrett began putting together a defense to Lillian Knox's grand larceny and perjury charges, he served as both Lillian's attorney and investigator because she didn't have the funds to hire an independent investigator.

Barrett's first step was to replace the public defender assigned to Lillian. He did so with a court appearance on March 26, 1957. In the same appearance, he asked and won the right to see the State's Attorney's list of witnesses who might testify against Lillian. The court set Lillian's case for trial two months later.

Barrett needed the two months to determine the state's strategy. As he waited for the list of witnesses, he requested a conference with his opposing attorney, which took place in the offices of State's Attorney Benjamin S. Adamowski in the Chicago Criminal Court Building. Here, he came face to face with Wilhelm J. Levander, who would prosecute Lillian. "He seemed somewhat uninterested in the case, probably because he had an excessive case load," said Barrett. "He also appeared to be unfamiliar with the facts of Lillian's case."

Levander, however, did give Barrett a list of his witnesses. The list included the Moir estate attorneys, Herman Fischer and J. Milton Guy; George Groethe, auditor of the First National Bank; Mrs. Irene Montgomery, Rose Moir's niece; her physician, Dr. Vincent V. Marzano; Chicago police officer Helen McGill; and the name of a mystery man, James Reddin. Barrett wasn't required by law to provide Levander with a list of defense witnesses.

In another meeting, this time with the Public Defender's Office, Barrett found the attorneys happy to be relieved of the Knox defense. They confirmed they had advised Lillian to plead guilty to both the grand larceny and perjury indictments, even though Barrett pointed out that a perjury conviction was at the time non-probationable. If Lillian had admitted she was guilty, she would have gone straight to prison for the second time in her life.

Now, Barrett turned to the process of interviewing the state's witnesses and finding witnesses who would be friendly to Lillian.

Lillian's friends, Pat O'Connor and Lola McElroy, provided Barrett with the names of people who might be able to shed additional light on the Moir-Knox relationship. Barrett wanted to know if they were good friends or just an employer and employee. Was Rose Moir likely to give large monetary gifts to someone like Lillian? Did anyone see or hear her make the gifts? Was she mentally competent to do so?

When Barrett interviewed Ethel Turner, who had been Mrs. Moir's maid at her last residence, he found her friendly and happy to talk. Ethel, a large black woman, was employed as a maid by Rose when she lived at the Bryn Mawr Apartment Hotel from January 9, 1954, until her death in December of that year. She said Mrs. Moir called Lillian "to her side maybe once or twice a week to kiss her on the forehead" and tell her, "I appreciate what you're doing for me, Costello (Lillian's alias)."

Ethel said Lillian bathed Rose, administered her medicine, groomed her hair and nails, gave her facials, massaged her with witch hazel alcohol, and cleaned Mrs. Moir after bowel movements. "Mrs. Costello didn't complain at all, and whatever Moir gave her, Knox was worth it. Knox had big circles under her eyes from taking care of her," Ethel said.

Ethel also reported that a month before she died, Rose gave Lillian a box and told her: "Costello, I have a Christmas present for you, and handed the box to Mrs. Costello, who took the box and thanked Rose and kissed her on the forehead or cheek. Mrs. Moir then looked at me and winked. Mrs. Knox didn't open the box."

Lola McElroy came to see Mrs. Moir every Saturday, Ethel said. "On the day that Costello went to the bank, I heard Mrs. Moir telephone Miss McElroy. I remember Costello leaving for the bank, and returning alone with a briefcase. It looked full and fat. Mrs. Moir laid it on the bed beside her."

Ethel said Harry Moir, Rose's adopted son, "used to come in quite often for money; I always knew he had been here because he would leave three or four empty whiskey bottles in the hall." Ethel said Herman Fischer also called on Rose once or twice a week. "Mrs. Moir often told me that Mr. Fischer had her put in that asylum because he wanted the key to her safe deposit box, and she wouldn't give it to him. She'd get really mad."

With Ethel's interview, Barrett had corroborated Lillian's contention that Rose had given her the box containing the bearer bonds.

From Irwin Hayden, a Radio Flash taxi driver and ordained minister who became friendly with Mrs. Moir, Barrett again confirmed the cordial relationship Lillian and Rose enjoyed. He said Mrs. Moir was alert and aware in her environment. "She appeared happy, talkative, attentive and friendly," he said. "My impression was, though she was in bed, she was alert and content," he added. Hayden also reported that on one occasion he had driven Lola McElroy to the bank, where she met Lillian. On another occasion, he drove both of them to the bank. "They told me they were doing some business for Mrs. Moir at her safe deposit box. There was quite a discussion about her will. Mrs. Moir had told them she was going to make out a new will," he said.

Barrett also found Lola to be a helpful witness when he talked to her. She also confirmed the friendship of Lillian and Rose. She said she was present in Rose's apartment when it was sealed after her death. "There were no stocks or bonds in evidence. The owner (Vincent V. Marzano) took us (Irene Montgomery and Lola) into his office and told us of some stocks and bonds which he said he found in the apartment," she said.

Dr. S.G. Schmidt, another Moir physician, said the last time he had examined Rose (before her death), "she was as mentally alert as any person her age could be." Dr. Schmidt said he was not aware of Mrs. Moir's financial background or Mrs. Knox's relationship with the deceased. Dr. Schmidt said if he had known of her financial circumstances, he would have charged her more. Barrett decided the doctor would not make a good witness.

Barrett also interviewed Nat Saletko, a Railway Express

employee, by telephone. He, too, said Lillian appeared to take good care of Rose.

"It was beginning to dawn on me that what took place in the New York police station in October or November of 1956 might be crucial to Lillian's defense," said Barrett. "I was beginning to wonder how the state planned to prove its case. What if neither of the New York cops who first arrested Lillian didn't show up to testify? Would the so-called 'chain of evidence' be broken? I had to talk to Officer McGill," he concluded.

When he cornered McGill in the Criminal Court Building on May 27, 1957—the day originally set for Lillian's trial—he found her to be slim, attractive, friendly and with no prejudices in the case. She answered all of his questions without hesitation.

McGill retraced her steps on October 29, 1956, when she was assigned to pick up Lillian Knox and the bearer bonds. When she identified herself to the New York Police Station as a Chicago policewoman assigned to retrieve Lillian and the bonds, two New York officers led her to a room containing a large safe. One opened the safe and retrieved a large packet of documents. Together, the three went through the papers—a collection of bonds from different states totaling about $50,000, none of which bore the names of owners. Responding to Barrett's questions, McGill said she checked the bonds against a list given her in Chicago. Lillian's name did not appear on any of the bonds.

With the inspection completed, McGill gave the New York officers a signed receipt for the bonds, put them in her briefcase, and walked out of the room with the officers and the executed warrants. When she went to the station's main room to pick up her suspect, she was approached by the two New York policemen who had arrested Lillian. They started laughing, and told McGill they had traced Lillian's address through some of the bonds' interest coupons, which had been cashed in New York. They went to the Woodstock Hotel and discovered her in Room 626, living with a 23-year-old Greek sailor who couldn't speak English. Found to be an illegal resident of the U.S. who had jumped ship in New York, the sailor was deported.

When McGill identified the two policemen, Barrett found his mystery man, James Reddin, one of the two arresting officers. The other was an officer named Arrow.

Reddin and Arrow said Lillian had freely admitted having the bonds and led them to a locker in Grand Central Station, where she surrendered the documents to the policemen. Reddin and Arrow told McGill that Lillian had "given them a cock-and-bull story about how the bonds had been a secret gift from a Mrs. Moir of Chicago, and that she wasn't aware of the bonds until much later." The officers thought her explanation was "pretty clever, if far fetched," said McGill, who wound up her conversation with Reddin and Arrow by reminding them that one of them would probably be called to testify at Lillian's trial in Chicago. She then left to find Lillian.

"I was shown to the women's lockup. There I was introduced to a heavy-set older woman, probably in her seventies, about five feet and three inches tall, with hennaed gray hair. She told me she was Lillian Knox. She was very friendly, even ingratiating, and had this very crafty, cunning look in her eyes. I knew she was sizing me up. She was pleasant enough, and I figured she wouldn't give me any trouble," said McGill.

With her briefcase containing the bonds under her right arm, McGill produced her handcuffs , snapping one on Lillian's right wrist and the other on her own left wrist. She signed out, the lockup door was opened, and McGill and Lillian entered a police car which carried them to La Guardia Airport.

On the way to the airport, on the flight to Chicago, and in a car to the Chicago police station, the bonds remained with McGill, separated from Lillian by the policewomen's body. Lillian never touched the bonds, never said anything about them, and never said how they came into her possession. Barrett then asked a crucial question: "Did you ever see the bonds in Mrs. Knox's possession?"

McGill's answer was no—and Barrett had his defense strategy. And he wouldn't have to put Lillian on the stand to be subjected to a withering barrage of questions from the prosecuting attorney and the damaging disclosure of her old prison term, perhaps even a resurrection of her shadowy life in Texas.

17

FACING A JURY. After his conversations with Officer McGill and Lillian Knox, Edward Barrett heaved a sign of relief when he received a notice from Assistant State's Attorney Wilhelm F. Levander that on May 27, 1957—the date set for Lillian's trial—Levander would appear before trial Judge James R. Bryant and request a continuance (delay) of the trial until June 17 on grounds that the prosecution would be unable to produce all of its witnesses on the scheduled trial date.

"The delay was a godsend, enabling me to do a thorough research job and to prepare some motions I had in mind," said Barrett. On May 27, he appeared in court with Lillian. Over Barrett's mild protest, Judge Bryant granted a continuance to June 17, but he also remarked: "This lady has been sitting in jail too long; no more continuances."

Barrett then turned to his partner, Rhea Campbell, a resourceful researcher into relevant cases, to find appellate or Supreme Court decisions which would support a strategy for extracting Lillian from jail. What he and Campbell were looking for were cases which held that the trial judge had erroneously admitted evidence prejudicial to the defendant.

They hit pay dirt in a few days when Campbell dug up a 1924 case, *People v. Maruda*, in which the defendant had been convicted of stealing merchandise from Marshall Field & Company, where he was a night porter. A number of dresses, a watch, a handbag and ladies hosiery were found in his home. Witnesses testified that some of the dresses and hosiery were from the Marshall Field Chicago store, but they could not say that they had been stolen. The Illinois Supreme Court reversed the conviction, holding that the prosecution had not proven the necessary *corpus delicti,* i.e. that a crime had been committed. The Supreme Court said, in part, "...the mere fact goods were missing, with no proof of circumstances, does not prove a larceny. It must be shown that their taking was without the knowledge or consent of the owner..."

When Barrett and Campbell found supporting cases in two other conviction reversals, they had good reason to believe they could win Lillian's case, especially if the New York policeman, James Reddin, failed to appear at the trial. He was the only person who could connect Lillian to possession of the Moir bonds.

"With these precedents and other cases, I prepared a motion for directed verdict at the close of state's evidence, which would be presented at the trial when the prosecution rested its case. If this motion was successful, the court would instruct the jury to acquit Lillian on the grounds that the state had failed to produce a *prima facie* criminal case, or had introduced evidence so weak that the defendant was not even required to rebut it," said Barrett.

While this strategy could remove the larceny indictment against Lillian, it would not quash the perjury indictment. "But I was confident that if Lillian was acquitted of stealing the bonds, she could not be tried for lying about them," Barrett recalled.

Meanwhile, in the event his directed verdict motion might be denied, Barrett prepared Lillian for a cross-examination. He also instructed her to play and dress the part of a nice little old grandmother, instructing her to let her hair revert to its natural gray.

"I warned her that if she had to take the witness stand, I would be the first to examine her, that she should not volunteer any answers, and that she should answer the prosecutor's questions on cross-examination as pleasantly and demurely as possible," said Barrett. Lillian, who had probably been in more criminal hearings than Barrett, said she understood and promised to behave.

Barrett was also ready to question Helen McGill and was prepared to put Lola McElroy and Ethel Turner on the stand as witnesses to the fact that Rose Moir had a genuine affection for Lillian.

On June 17, a rainy and cool Monday in Chicago, Barrett showed up an hour early for the 9:30 a.m. trial in the Cook County Criminal Courthouse at the corner of 26th Street and California Avenue. He had driven from his home in Winnetka, twenty-five miles north of Chicago. He went to Judge Bryant's chambers and found Lillian waiting patiently in the anteroom, guarded by a woman police officer. She was not wearing handcuffs and, at Barrett's suggestion, had dressed the role of a grandmother. Her hair was now gray and she was wearing a plain dress furnished by Lola McElroy.

Barrett explained what would happen during the proceedings. He said the attorneys on both sides would probably spend some time interrogating prospective jurors as to their qualifications, backgrounds, and prejudices—and, in the end, rejecting some and accepting some. Once a jury was chosen, opening statements would be made by the opposing lawyers, followed by the presentation of witnesses and evidence. "I was really hoping for a directed verdict because I feared the jury wouldn't believe Lillian's story," said Barrett.

When Judge Bryant—noted for his promptness, meticulous manner and even-handedness—opened the trial, Barrett presented a motion to exclude all prospective witnesses from the courtroom so that the prosecution's witnesses couldn't listen to the other witnesses. Judge Bryant granted the motion.

Jury selection went quickly and Barrett and Levander picked a panel of four men and eight women. All were middle-aged and blue collar, including one black woman. Two had previously served on criminal juries. At 1:30 p.m., Judge Bryant called a recess, giving Barrett a chance to remind Lillian what he expected from her testimony before she was taken back to the jail. He also complimented her on her demure demeanor.

When the trial resumed after lunch, Barrett was faced with the decision of making an opening statement. "I reasoned that I might talk my defense into a Catch 22. The only thing I could tell the jury was that Lillian would testify about Rose Moir's secret gift. If I did so, I would supply the 'possession link' in the chain of evidence which would be missing if policeman Reddin didn't testify. A defense attorney can bind his client by his own words. I couldn't take that chance. If the New York cops didn't show up, my opening statement might cook Lillian's goose by my own admission," recalled Barrett.

When Judge Bryant called the court into session and ordered the jury seated, Barrett went to the anteroom and escorted Lillian—by now a demure little old lady—to her seat. She walked lamely, leaning on Barrett's arm.

When the judge called for opening statements, Levander stood up, walked to the jury box, and told the panel the case involved the prosecution of Lillian for the grand larceny of $53,000 in bearer bonds belonging to the estate of the late Rose Moir. Levander said Lillian had testified in probate court that she didn't have the bonds, but had been arrested with the bonds in her possession in New York. Levander said she took the bonds with felonious intent. When Levander finished, Barrett rose and said he would reserve his opening statement until the close of the prosecution's case.

Levander then moved ahead with his list of witnesses, beginning with Ralph Youngsma, who testified that he was the assistant cashier and operating manager of First National Bank's bond department, that he knew Mrs. Moir 25 years, and that the $53,000 in bonds found in New York were listed as having been sold to Rose. In response to a single question from Barrett, Youngsma said that Mrs. Moir owned the bonds in December of 1954, as far as he knew.

The next witness was the Moir estate attorney Herman Fischer, who in a rambling discourse described his acquaintance with Mrs. Moir and her late husband since 1915, when he began representing the Moirs' interests. In early 1952, he took Rose from the Morrison Hotel and placed her in the North Shore Mental Hospital, where he visited her every week. Thereafter, she resided in a series of hotels, ending up at the Bryn Mawr, where she died in December of 1954. He said that following Rose's death, Lillian gave him more than $200,000 in stocks belonging to Mrs. Moir. Fischer also identified the bearer bonds and said he was the administrator to collect, as approved by the probate court. Under cross examination by Barrett, Fischer admitted that Mrs. Moir was never adjudicated mentally ill. He also admitted that he didn't know for sure whether Rose owned the bonds at the time of her death. Dr. Marzano and the owner of the Bryn Mawr, testified to the same effect. "All Dr. Marzano could contribute was that Rose was physically weak the two times he saw her and that he found a two and a half foot by three foot trunk in her living room. On cross examination, I brought out that he never attended Mrs. Moir as her physician. His input was negligible," said Barrett.

George A. Groethe, the bank's auditor, said he examined the bearer bonds and confirmed they were listed as those sold to Mrs. Moir. On questioning by Barrett, Goethe also admitted he didn't know whether Mrs. Moir owned the bonds at the time of her death.

When police officer Helen McGill was called to testify, she told substantially the same story as she had earlier given Barrett. She said she had a conversation with New York policemen Reddin and Arrow in New York. On Barrett's objection, she was not permitted to disclose the contents of the conversation because it was considered hearsay. On cross examination, McGill said she had never seen Lillian in possession of the bonds and that Lillian never admitted possessing them. As Barrett expected, McGill never connected Lillian with possession of the bonds.

When Levander rested his case at the end of the day, Barrett sighed with relief. Policeman Reddin did not appear—and the chain of evidence had been broken.

When the trial resumed the next morning, Barrett approached Judge Bryant's bench and presented his "Motion for Directed Verdict," a type of motion presented in most jury trials, and usually denied. The motion read: "Now comes the defendant, Lillian Knox, by Edward J. Barrett and R. Douglas Campbell, her attorneys, at the close of all evidence for the State and moves the Court to instruct the jury to find the defendant, Lillian Knox, not guilty." The motion was accompanied by a brief citing the *Maruda* case and others. Barrett argued that Levander had failed to produce a *prima facie* case and that Lillian wasn't required to mount a defense.

Judge Bryant agreed. He quickly granted the motion, called the jury back into the courtroom, and instructed them to return a verdict of "not guilty." The bewildered jury complied. In a two-day trial, Barrett never had to put on a defense, Lillian never testified, and in one stunning moment, Texas' Lady Bountiful was freed from another indictment hanging over her head. Although she had faced a jury for the first time in a series of court cases spanning thirty-five years, she had not been placed on the stand to testify or faced questioning by the prosecution.

Barrett then began work on Lillian's perjury indictment, which alleged she had lied about removing the bearer bonds from the Moir apartment. On June 25,1957, he petitioned Judge Bryant to release Lillian on her own recognizance, pending a hearing on the motion to dismiss the perjury indictment. Judge Bryant granted the motion and Lillian walked out of the Chicago jail after six months behind bars, returned to Lola McElroy's Castle Hotel, and assumed her alias of Margaret Costello.

Three days later, Barrett appeared before Judge Dunne in

Probate Court as Margaret Costello's attorney and asked for a dismissal of the perjury indictment. On July 19, Dunne dismissed the indictment, putting Lillian/Margaret beyond the reaches of the Moir estate lawyers. She could now claim with impunity that she had possessed the bearer bonds and could claim ownership.

On July 23, 1957, at Lillian's request, Barrett wrote John W. Minton, her old Hemphill, Texas, lawyer, to keep him "abreast of developments in her various legal entanglements." But Barrett wasn't quite finished with Lillian's legal entanglements; he still had some unfinished business with the Moir estate's lawyers.

On July 23, he had filed an answer to the estate's rule to show cause why Lillian should not be held in contempt for failing to deposit a sum of money she had agreed to deposit with the Probate Court in 1954 when she fled to New York. He also filed a series of other motions, including a petition to vacate and set aside a Probate Court order entered in November of 1956 declaring the bearer bonds to be the property of the estate. In addition, he filed an answer to the rule to show cause, alleging Lillian had no knowledge of the entry of the order entered in 1955.

On July 25, Judge Dunne struck the answer, but allowed Lillian to remain free on her own recognizance. Barrett quickly filed a petition to vacate Judge Dunne's order of November 24, 1956, which had declared the estate to be the owner of the bonds. On August 29, Judge Dunne struck the petition, but allowed Barrett to file an amended petition. In September, his amended petition set forth Lillian's "secret gift" allegations, admitting for the first time she had possession of the bonds she had been acquitted of stealing. The Rose Moir estate responded with a motion to strike the amended petition and on December 18 Judge

Dunne granted the estate's motion, ordering the State's Attorney to release the bonds to the estate. On January 10, 1958, Barrett filed a notice to appeal Judge Dunne's decision to the Cook County Circuit Court.

As he waited for the Circuit Court to make a decision, Barrett felt Margaret had "one more string to her bow," as he described it. He believed Margaret had sufficient grounds for filing a lawsuit against Herman A. Fischer for malicious abuse of process and malicious prosecution. However, when he discussed the subject with Margaret, she did not seem overly interested.

In June, 1959, as the statute of limitations for such a lawsuit neared its deadline, he prepared a lawsuit to be filed against Fischer in Cook County's Circuit Court, seeking damages for $150,000, and wrote Margaret at the Majestic Hotel in Chicago, where he believed she was living. She never answered his letter, and the lawsuit was never filed.

Margaret, meanwhile, was obsessed with retrieving her meager personal effects which had been in storage since 1955, when they were attached by order of the Probate Court. Barrett told her it was probably a lost cause, but promised to help. He wrote letters to Reebie Storage Company, to Margaret's previous attorney, David Fisher; and to the Moir estate. Fisher did not reply, but J. Milton Guy, a lawyer for the estate, sent Barrett on May 10, 1960, copies of a court order directing the sale of the possessions. The storage bin included municipal bond coupons from Montana, Illinois, New York, California, and Minnesota valued at $413.89; two coin purses containing $10.90, and a collection of costume jewelry, women's clothing and purses, household equipment, and various books and, cosmetics, small pieces of furniture, papers and personal items.

Barrett learned on March 2, 1960, that Reebie had distributed most of the belongings to Fischer and they were sold by order of the court. Irving Gold, a Superior Court deputy, had bought the items, presumably for resale. Barrett then advised Lillian of the sale and told her "you therefore have no legal standing to complain." Barrett, however, managed to retrieve from Fisher and Fischer some papers, mostly chattels and correspondence, belonging to Lillian.

On June 30, 1959, after a year and a half of continuances, Judge William J. Touhy, in a *de novo* hearing (complete re-trial), vacated Judge Dunne's dismissal and returned the case to Probate Court.

After another flurry of motions and legal moves, both sides in the dispute decided to seek a settlement, to Judge Dunne's relief. On November 2, 1959, Judge Dunne called a conference of the attorneys. In another meeting in December, the estate offered $5,000 to Lillian.

Assuring Judge Dunne that he would submit the offer to Lillian, Barrett tried to locate his client. But she had disappeared again.

18

LILLIAN DISAPPEARS. Since leaving a federal prison in 1944, Lillian Knox had shrouded her life and her background in secrecy. Using a variety of aliases, she avoided her real identity. She seldom made personal telephone calls and directed her friends and relatives to mail her letters though intermediaries.

Now, in the winter of 1959, her attorney, Edward Barrett, was confronted with Lillian's penchant for secrecy—an idiosyncrasy that could cost her thousands of dollars during a time when she was virtually penniless.

He had won for her a $5,000 settlement offer in the Cook County Probate Court, but he was unable to reach her by telephone. She had left her current residence and moved to parts unknown.

He wrote a letter, addressed to Margaret Costello, and sent it to a woman living near San Augustine, Texas, identified by Lillian as her sister, but received no reply. On January 21, 1960, he wrote a second letter to Margaret, addressed to Henrietta Green of Hemphill, a name mentioned by Lillian in a conversation. He advised her that attorneys for the Moir estate had agreed to pay $5,000 to settle the issue in Chicago. "Mrs. Costello, the choice is up to you, but I believe that you would be well advised to make a settlement at this time. I believe Judge Dunne is going to rule against us in the event we do not settle, and if this happens, we will have to appeal to the Appellate Court this time, with attendant expenses and a great expenditure of time..."

But Barrett's second letter also went unanswered and in February of 1960, in another conference in the chambers of Judge Dunne, he said he couldn't accept or reject the $5,000 offer because he couldn't find Lillian. Judge Dunne then persuaded the Moir estate to increase the offer to $10,000, and gave Barrett ten days to find Lillian and obtain her acceptance, or the settlement offer would be revoked.

In desperation, Barrett called the law offices of John W. Minton, Lillian's old attorney and confidant in Hemphill, Texas. Minton had died in 1959, but Barrett talked to his son, John Henry Minton, also an attorney. He said Lillian was not in town, but urged Barrett to come to Hemphill and said he would attempt to help. Barrett flew to Shreveport, Louisiana, rented a car and drove seventy miles to Hemphill, just across the Sabine River from Louisiana.

In mid-February, Barrett arrived at the Hemphill courthouse square—where a Sabine County grand jury in 1923 had no-billed Lillian on charges she murdered her husband Hiram. After making an appointment to meet Minton at 6 p.m. at his home, he strolled around the square and ordered the blue plate special at a local cafe.

At 6 p.m., he arrived at Minton's home west of Hemphill. "He was dressed in a business suit, and looked every inch the prosperous local lawyer," recalled Barrett. After a few minutes of conversation, Minton's tone turned cordial and helpful. He directed Barrett to drive to "the colored part of town" and on their way, he reminisced about Lillian and Hiram. They soon arrived at an old sawmill shack, where Henrietta Green lived. Minton said she was a widow who had been one of Lillian's servants in the Knox mansion. She, he said, might know Lillian's whereabouts.

"Mrs. Green was a short, older widow lady, quite black, age about seventy, I guessed, and was wearing some kind of flowered house dress and a bandana on her head. Her house was a one-room shack with a dirt floor. There was a bed, two chairs, a stove and very little else. She had a religious calendar and a few religious pictures on the wall. It was Spartan, but clean and warm. She sat on the bed; we sat on the chairs. She was pleasant, but guarded. I was sure she would not have talked to me if Mr. Minton had not been there," said Barrett. "At first, she didn't say anything, but just looked me over, so I looked her over, too."

In looking around, Barrett noticed cartons of canned goods in a corner of the shack—an obvious link with Lillian and a sign that she still had a soft place in her heart for her old employees, particularly those with limited resources.

Minton convinced Henrietta that she should, for Lillian's sake, talk to Barrett. "Finally, she told us she loved Lillian for all she had done for her and her friends and relatives in Hemphill. She indicated that Lillian was using her as a mail drop. All letters sent to Lillian were addressed to Henrietta, who would then mail them unopened to Chicago in pre-addressed, stamped envelopes. Lillian had taken great care to prevent anyone from finding her," said Barrett.

When Henrietta gave him Lillian's forwarding address at 3757 West Giddings Street in Chicago, Barrett realized he had come nearly a thousand miles to Texas to discover Lillian was still in Chicago.

Thanking Henrietta for her help, Barrett and Minton left. Minton had arranged for Barrett to spend the night in San Augustine at Mrs. Brown's boarding house, which was comfortable, but lacked hot water. The next morning, with time to kill before his late afternoon flight to Chicago, Barrett rejoined Minton in Hemphill. Together, they drove around Hemphill while Minton retold the story about Lillian and Hiram's sawmill, their ascendancy as millionaires, and the death of Hiram on November 26, 1922—which started Lillian's series of confrontations with the law.

On the return trip to the Shreveport airport, Barrett made a detour to Fairway Farms Club, a golf course. He introduced himself to the manager, explained why he had been in Hemphill, and was invited to lunch. In the dining room, he saw a table surrounded by middle-aged women, introduced himself, and explained that he was trying to locate Lillian Knox. "When I said her name, all of the women perked up and immediately recognized the name; they couldn't wait to tell me what they knew about Lillian, most of which came from stories passed down by their parents. They told me about her scandalous parties, her notorious largess, and the murders of Hiram and his mother, Mary Knox, in Dallas. Lillian, I found out, was a *bona fide* Texas legend," said Barrett.

Returning to Chicago in early evening, Barrett drove to the Giddings Avenue address, a typical three-flat apartment on the

Northwest side of town. He rang three doorbells before he located Lillian, now back to using the alias of Margaret Costello, living in a first-floor apartment with a middle-aged couple, Bernard and Esther Gamm. Lillian was working as a practical nurse, as she did with Rose Moir. As Barrett looked around the apartment, he saw dozens of cartons of canned goods stacked to the ceiling. Lillian had not given up her trademark confidence game.

Barrett chided Lillian for not answering his letters and leading him on a wild goose chase. With only a couple of days left until Judge Dunne's deadline, he explained to Lillian the urgency of agreeing to the probate court settlement. She reluctantly signed a release and the authorization empowering Barrett to accept the $10,000.

"I explained to her she could take a lot of satisfaction in turning the tables by actually getting $10,000 from the people who had accused her of larceny and perjury and had her jailed. But I got the feeling she didn't covet the money as such. It was more like a game to be won, and dollars were how you kept score," said Barrett.

On March 15, 1960, Barrett distributed the money to Lillian, minus his legal fees and costs. She immediately deposited the remainder in his law firm's trust account and asked him to hold the funds for her. She said she would withdraw small amounts from time to time.

Lillian left Barrett's office and returned to the Gamms' apartment, but she stayed in touch with him for the next two and a half years, living mostly in cheap hotels and occasionally requesting small amounts, usually $100, from the trust account.

On April 25, 1961, Lillian showed up at Barrett's office to

discuss with him the status of old oil lease lands which she claimed she owned. She told Barrett this story: "The oil lease lands I got from my husband just before his death in 1922 were in two parcels. Parcel one consisted of 10,402 acres leased to Gulf Petroleum Company in Houston. Parcel two consisted of about 6,500 acres leased to Magnolia Oil Company of Shreveport. Both were ten-year paid-up leases at thirty to fifty cents per acre. The leases were paid up in advance with options to buy after the ten year terms," she said.

On April 26, 1961, Barrett wrote identical letters to Gulf Petroleum Company and Magnolia Oil Company, seeking information on the leases. Magnolia's successor, Mobil Oil Company, replied that it did not hold any oil lease covering 6,500 acres near Hemphill or in Sabine County. Gulf said it held at one time a lease on Knox properties, but did not currently have such a lease. Barrett notified Margaret and the oil lease issue was dropped.

"Next, Lillian wanted to sue Chester B. Collins, her former attorney from Texas and, according to her, an ex-husband, but she couldn't define the grounds for the lawsuit, so the issue was dropped," said Barrett. In a letter to Barrett on February 28, 1960, Lillian claimed Chester was fearful of federal courts and said "he will settle, but not until the suit is filed and laid on his desk. It will have to come as a surprise, a complete surprise." She added: "He tried to kill me twice; he thought he had once. Mr. Minton tried to get me to sue him." Collins died had in Fort Worth on October 20, 1960, at the age of seventy-two.

Lillian's growing paranoia may be attributable to a disease which had begun to affect her cognitive abilities. Sometime in the early 1960s, she made a trip from Chicago to Houston for an appointment with Dr. Hugh Arnold of Houston, a son of the Knox family physician in Hemphill, Dr. Tom Arnold. Lillian may have known Hugh—one of the founders of Houston Diagnostic Clinic—because of his father's past association with the family, or it may have simply been an ironic coincidence that she arrived in his examining room.

Before his death, Dr. Arnold described her visit to his family. He said she arrived wearing a heavy coat and did not remove the garment when she was weighed. Since her weight appeared to be excessive, she was asked to leave her coat with an attendant during the medical examination. While hanging up the garment, the attendant noticed that a large amount of papers—which appeared to be deeds and other legal documents— pinned inside the coat. She was carrying with her years of accumulated documents and papers that should have been kept in a safe deposit box or another secure place.

Dr. Arnold ordered blood tests for Lillian and examined her. He concluded her mind was affected by the disease she bore. Explaining this to Lillian's son, Hiram, and his wife, Dr. Arnold said they appeared to be afraid of Lillian. They explained that, during her stay in their Houston home, they took turns staying awake because they were concerned by Lillian's odd behavior.

Two separate visits Lillian made to her children in Texas between the mid-fifties and 1963 illustrate the progressive deterioration of her mental capacities.

In the mid-fifties, she visited with her sisters, Evelyn Troup and Fannie Watson, on a ranch near Austin and with her son, Merton (Bud), at Schertz. During that visit, she was alert, normal and pleasant. She returned to Chicago after giving all of Merton's children gifts of money.

But in 1963, on a return visit to Merton's family in San Antonio, she acted "really weird," said grandson Mark Knox. "She locked herself in my room, eating peanut butter and watermelon, frequently spitting the seeds on the wall, where they stuck. She also stuffed some of the watermelon seeds in a large pair of ladies panties and hid them behind my dresser," said Mark.

During the visit, Merton received several crates containing household items from Alaska, where he had completed a tour of duty with the U.S. Air Force. As he cleaned out the crates, he threw some damaged clothing and other unwanted items into a garbage can. Watching him, Lillian became agitated and called Randolph Air Force Base in San Antonio, where Merton was assigned, to complain that "my son is stealing my stuff."

"When she left, she asked Dad for bus fare money for the return trip to Chicago," said Cindy Knox Fuller, Merton's daughter. "But when my brother and a friend took her to the bus station, she pulled a wad of money out of her bra, and gave them more money than the bus fare had cost her," she said.

In the sixties, two of Lillian's sons, Marcene and William—feeling the Knox children might have some ownership rights to the Knox timberlands in East Texas, as well as possible lands in Oklahoma— discussed a possible lawsuit or a title search. But nothing came of the effort.

19

A PAUPER'S GRAVE. After 1961, Edward Barrett did not see Lillian Knox as frequently as he did in the years before. She only visited his office when she needed small sums of money from the trust account managed for her by Barrett's law office.

On one of her visits, she divulged—in an apparent break from her tendency to keep her whereabouts secret—that she had moved out of the Gamms' apartment and was living in a hotel. Each time Barrett saw her after that, she was living in a different hotel.

"On one occasion, I went to see her in a run-down residential hotel in Chicago's uptown district on the north side. I found her alone in a pathetic little room, cooking on a hot plate. The room had an overpowering odor of grease. She was heavier than I had ever seen her, and she had difficulty in moving around the room. She must have been eighty years old, although I never knew her exact age, but she looked like it," said Barrett.

He was appalled at seeing her in such dire circumstances

but, as always, she was in good spirits, happy to see her old lawyer friend, and full of conversation.

"I asked about her friends and wondered if she ever saw them. She said that Lola McElroy and Pat O'Connor still came around to see her every now and then," said Barrett. "But, knowing of her penchant for secrecy, I wondered if she was being honest with me," he said.

Lillian had carried on an infrequent correspondence with Barrett in the late 1950's and the early 1960's, sending letters to him from different addresses and always moving around.

In 1959, she wrote him: "Some three months ago, my sister had an operation for cancer (which had apparently returned after a remission). She has been getting worse the last month. I received a telegram and so I am leaving. I have checked out of the hotel. They are also moving the elderly woman (apparently someone Lillian was nursing) to her nieces in Skokie tonight, so I am going out with them. I did not know this when I called you. My sister lives on a farm 42 miles on a highway from San Augustine (however, she may have meant San Antonio). She has no telephone, the nearest telephone is 12 miles (away). You can only reach me by letter. I would have to have at least two weeks notice to come back for court. Please forgive me. I could not go through with the publicity, which I know would happen. If I had money where I could pay room rent and get food, I would go along, but I was almost starving for over one and a half years. Please, please understand and forgive me. I will do my best to repay you soon. Please know it (is because) of the lack of money. I have not yet paid the money I borrowed from salaried folks (and) borrowed from you. I have and will continue to pray for you and Mr. Campbell's health, and for God to give you both his richest blessing. So long as I live, you both have my prayers. Gratefully. Please do not give anyone my address. If you can, just say I am out of town..."

On January 22, 1961, she wrote Barrett again, requesting $200, and adding: "If you can not telephone me (meet me) at Lane Hotel Whitehall, room 415...If I do not hear from you I will come up about 2:30 p.m."

In another letter, this one undated, Lillian told Barrett, "A man is going to Beaumont, so I am going to give him this letter to mail...it will reach you sooner. I am not in Beaumont, I can not be reached by telephone. I can get Mr. O'Connor...I can go and call him station-to-station to a town where I am staying, not (in) Hemphill. I can not go at night to telephone you unless the weather gets better. See if you can get a $5,000 fee for your firm, then you can also take out of what ever else we get. You both have worked so hard. I have no way to give or get you any money. My sister in Beaumont does not know it...she had a small cancer on the side of her nose. They, the doctors, feel they can cure it. I should go to see her, but I don't have any money and none to help her. Not having money, being ill for medicine and food, I had to come where I could get it. They are not any kin to me. Being of service to these fine people when we had money, they treat me with welcome..."

From time to time, she would show up unannounced at Barrett's office. "We knew what she was, but we were captivated by her personality and her engaging rascality. It was an event when she showed up," he said.

On August 21, 1962, Lillian arrived at Barrett's office for the last time to collect the balance of the funds he had been holding for her. He invited her into his office and they began to reminisce about their experiences.

Using her alias, as he had always done, Barrett told her: "Margaret, you've never told me, but I'm sure you must have lots of money hidden away somewhere, maybe in out-of-state banks under assumed names. Why don't you use some of it to live decently, instead of staying in flea-bag hotels?"

Lillian said nothing, but gave Barrett a wise, knowing look that he knew so well.

Barrett added: "If I'm right, and you have funds hidden somewhere, you wouldn't want them to be lost when you die. After all, you're not getting any younger. If you wish, we can set up a secret trust so that Mr. Campbell and I can have access to the funds after your death and distribute them to your children."

Lillian's enigmatic reply was, "Don't worry, Mr. Barrett. I've taken care of that already."

Barrett interpreted her reply as meaning that she probably had money hidden away somewhere, but knowing Lillian's earlier proclivities, it could have meant that she had already given it away to someone she felt needed money more than she did.

Lillian, he concluded, would probably continue to live as a pauper, far removed from her days as one of Texas' richest women. Barrett asked her if her children could help her in her old age. She bristled and said none of them could afford to help. One son, Merton (whom she called Bud) Knox, had cancer and a daughter, Mary Lyllian, was a Catholic nun known as Sister Mary Clement and had no money, she said.

Barrett then approached Lillian with a question that had been on his mind since her arrest in New York. "Margaret," he began, "I've got to admit I was surprised when I heard you were living with that young Greek sailor in New York. You were seventy-five and still interested in sex?" Lillian laughed. "Mr.

Barrett, you're only 39 years old and you're a man. What you men don't know is that a woman never loses interest in sex."

When they finished their conversation, Barrett never saw Lillian Knox again. Her closest Chicago friends, Lola McElroy and Pat O'Connor, passed away and Barrett stored Lillian's closed legal files in a cardboard box in his office. Lillian's old antagonists, Moir estate lawyers Herman A. Fischer and J. Milton Guy, had also died.

The newspaper headlines, court cases and rumors began to fade into history. But Lillian remained in the back of Barrett's memory. "I was attracted to her in a certain way because she had a magnetism about her. She could talk sincerely and confidentially with people, and had a way of winning them over," he recalled

In 1994, as he began recording his experiences with the one-time Texas timber baroness, Barrett learned that in April of 1965, while she was living alone in the North Meire Hotel in Chicago, Lillian became disoriented and unable to care for herself. An ambulance was summoned and Lillian, as Margaret Costello, was carried to the Kankakee State Hospital, about 35 miles south of Chicago, where she was admitted under her real name. The attending physician diagnosed her with "chronic brain syndrome associated with circulatory disturbance with cerebral arteriosclerosis with psychotic reaction." Her birthplace was listed as Texas and her age as 73

She died at 11:35 a.m. December 17, 1966, without any family members present. The death certificate attributed her demise to "cardiac failure due to arteriosclerotic heart disease. Her religion was listed as Catholic and her education as the 12th grade. Her file at the hospital listed three sons, Hiram of Houston, Merton of San Antonio and William, who was in a Texas tuberculosis hospital at the time, and one sister, Sister Mary Clement of Shreveport.

Lillian's oldest son, Hiram, was notified of her death. Asked if he would like to have her body transferred to Texas for burial, he declined. No members of the Knox family attended her burial by Rambo Funeral Home in grave Number 4608 in the Kankakee hospital cemetery. Merton Knox later learned of his mother's death from his brother and traveled from Dallas to Illinois to gather her belongings. He learned of two storage buildings she had rented in Skokie, Illinois, but found only old clothing and other items of little value. He gave the contents to the storage buildings' attendant and returned to Texas.

As the years passed, Lillian's grave was lost among more than 4,000 pauper graves at Kankakee. Even Marker No. 4608, the only link to her identity, has disappeared.

Since completing her prison term in West Virginia, Lillian apparently had only infrequent contacts with her children. But in her final years, she appeared to exhibit more pride in her offspring and tried to reach out to them with visits, letters and special gifts. Her children, however, were suspicious of her motives.

In a letter to Barrett, she wrote: "My oldest sons are coming to see me soon. When they come, I would love for you and Mr. Campbell to meet them. They are over six feet and handsome with nice personalities like you and Mr. Campbell...." In another letter, she said, "My son Marcene has entered a Catholic missionary college to become a priest. I have a beautiful daughter (who is) a nun. So God has blessed me..."

Lillian's children also benefitted from her odd practice of sending to her friends and relatives random collections of groceries, clothes and household goods. Grandson Mark Knox of Dallas recalled that when his father Merton and his family were stationed in Alaska with the Air Force between 1959 and 1963, Lillian sent the family "huge cases of items, such as large jars of petroleum jelly and peanut butter" from Illinois. She also sent purses, hats, old coats and sheets.

By the 1990s, all of Lillian's sisters and her brother were dead. Two sisters, Clara Marshall Blain and Evelyn Marshall Troup, lived in Beaumont, and a third sister, Fannie Marshall Watson lived in Houston and Austin. Her only brother, Alex Marshall, also lived in Beaumont.

By mid-year 2000, all of Lillian's six remaining children were also dead.Her youngest child, Mary Marguerite Signoria, had died as a teenager in 1939, after shooting herself while despondent over losing her boyfriend. Ironically, Mary had attended North Dallas High School with Dorothy Davis, who would later become the wife of Merton Knox.

William Hiram (Bill) Knox, a retired Air Force paratrooper and instructor at Fort Benning, Georgia, died at Houston on June 14, 1986. Bill was one of six Knox children born with a twin. His brother John died three months after the boys were born in 1912. Two other twins Lillian and Grace, who were born in 1913, supposedly died on the same day in 1914, but the details of their deaths are unknown.

L. Marcene (Mark) Knox also died in Houston in the 1980's. He had served with the U.S. Army during World War II and reportedly became the military mayor of Stuttgart, Germany, following the war. Mark joined the Catholic church, but never entered the priesthood, although his family said he talked about doing so. He was married to Mary Ellen Sawyer in July, 1943, but they were later divorced.

On January 25, 1988, Sister Mary Clement Knox, 70, who had been born as Mary Lyllian Knox, died at the Daughters of the Mother house in Shreveport, Louisiana, after a lengthy illness. Born at East Mayfield October 5, 1918, she had entered the Daughters of the Cross convent on August 15, 1938. She had taught elementary grades in a number of schools in North Louisiana and served as a principal of St. Matthew's School for two years. .

On September 7, 1991, Sister Mary Clement's twin brother, Merton, died in Dallas after a long battle with cancer. A retired U.S. Air Force lieutenant colonel, Knox had served as an Air Force pilot in the 17th Air Photo Reconnaissance Group, serving both in the Pacific theater during World War II and the Korean War. He received multiple decorations for his service, including the Purple Heart.

On July 12, 2000, Hiram William Ambrose Knox, the last of Lillian Knox's children and the last living link with the family, died in Auburn, California, one month short of his ninetieth birthday.

Even though he had faced a series of criminal charges during his youthful years in Texas and had been questioned in the death of his grandmother, Mary Knox, in 1937, Hiram had put his shadowy past behind him by becoming an engineer, working for large companies such as Brown & Root Construction Company and traveling to Norway, Germany, and other countries on construction assignments. He married Mary Elizabeth Burns of Big Springs, Texas.

Cecilia Fry, Hiram's niece in California, said, "Hiram had the mumps as a child, and they couldn't have children, but they became active in community groups and enjoyed each other," she said. "Hiram Knox was a self-made man, the salt of the earth, one of the finest men I have ever known," she added.

In 1991, Hiram Knox's family recorded on videotape some of his recollections about the Knox saga. When he talked about his youth in East Mayfield, he was lively and amusing and referred to Lillian as "mommy." But when he was asked about his mother's escapades in later years, he was hesitant and somber and referred to Lillian as "Mrs. Knox." When a relative asked him about Grandmother Mary Knox's 1937 death in Dallas, he said, "I don't want to go there."

Known as Curly because of his baldness, Hiram enjoyed practical jokes in much the same fashion as he did as a youngster in East Mayfield. While he was living in a nursing home in Auburn with his wife, a friend mentioned that he was planning a vacation to Hawaii and asked Hiram if he could bring him a gift. "Sure," said Hiram, "bring me a grass skirt and a coconut bra." When the friend brought back the items, Hiram stripped to his boxer shorts, donned the skirt and bra and did a hula dance in the nursing home.

But, despite his lively exterior, Hiram carried an inner disdain for his mother throughout much of his adult life. When he discussed her with family members and friends, he frequently said Lillian Knox "made my life a living hell" as a young man and tried to ruin the lives of his brothers and sisters. He said she tried to give away or sell some of her nine children, left one on the steps of a Catholic convent, and used the children to extract sympathy from people, even in her old age.

She said Lillian often called him an illegitimate child and said she didn't want him "gracing my door." According to a notation in the Knox family Bible, Hiram was born on August 7, 1911, but years later, he located his birth certificate, which indicated he was actually born in 1909—a year before Lillian had married into the Knox family.

Knox never forgave his mother for many of her misdeeds. One, in particularly, was the supposed theft of his Texas A&M memorabilia, including his Cadet uniform and boots, from a locker she allegedly looted while visiting his home. But he also found humor in the incident. "When Geraldo Rivera was opening the Al Capone vault on television in Chicago a few years ago, I expected to see some of our old Knox belongings," he said.

With the deaths of the Knox children, and the preceding deaths of the five Knox family members who came to Texas from Wisconsin around 1900, all ties to the family's Wisconsin relatives were broken.

Richard Hollar, a distant relative of the Knoxes, observed on August 19, 2002: "It is surprising to me that the Wisconsin relatives of Colonel W.H. Knox, namely the daughters of his brother, Samuel Knox, seem to have completely disconnected from their uncle after his move to Texas. Samuel and his brother were close associates for a long time and it appears their children grew up in Wausau up to a point."

Hollar observed that Samuel Knox's children became prominent in law and banking in the Wausau area and that his son-in-law, Andrew L. Krutzer, served as a Wisconsin State Senator. "The family was apparently proud of the Knox clan. Why they would have essentially forgotten or neglected such things as the remains of Colonel Knox in Texas escapes me," he said.

"I suspect that Lillian Knox had much to do with the matter at the time," he added.

Hiram W.A. Knox in a 1937 Dallas newspaper photo.

Lillian Knox is buried somewhere in the weed-choked cemetery of an Illinois mental hospital. *Photo, Edward J. Barrett.*

Hiram W.A. Knox in a 1991 videotape interview.

Epilogue

On June 10, 1922, Lillian Marshall Knox stood on an auditorium stage in Sabine County, Texas, to accept the accolades of an admiring audience. It was the finest night in the life of the woman known as Lady Bountiful—a generous person who cared deeply and compassionately for the poor sawmill families of East Texas.

On December 17, 1966, she was buried near Kankakee, Illinois, in a grave with only a number. Not a single mourner was present. She was poorer than the sawmill families she helped.

In the forty-four years that passed between her finest night and her last day on earth, Lillian Knox traveled down a tortuous trail of tragedies, plummeting from a millionaire's mansion to a penniless existence, from accolades as a gifted businesswoman to accusations as a murderess in two counties and imprisonment for mail fraud, and from the influence of a Texas timber baroness to the ignobility of stealing money from elderly.

On November 26, 2002, the 80th anniversary of her husband Hiram's death in Sabine County, there were few reminders of Hiram and Lillian and the multi-million dollar lumber empire they once owned.

The remnants included a Texas Historical Marker standing near the site of Knoxes' East Mayfield sawmill, two fragile albums containing photos of the Knox family taken between 1915 and 1920, the Knox family Bible with a family record in Lillianand Hiram's handwriting, a few tapestries and a rocking chair from the Knox mansion in East Mayfield, and an eight-foot religious oil painting, also from the Knox home, now hanging in the Shreveport convent where Lillian Knox gave up one of her daughters.

With these scant reminders, Lillian Knox remained as enigmatic in death as she did when she was alive. The questions which dogged the deaths of her husband—as well as his first wife Grace, his son Willie, and his mother Mary—continued to cast a black cloud over the Knox family. With the loss of evidence, public files and individual memories, answers do not appear forthcoming. Those who have studied the life of Texas' only timber baroness are usually drawn to one of two conclusions:

• Lillian Knox was a well-meaning, compassionate person who found herself caught up in a maelstrom of events that drew her deeper and deeper into a quagmire of rumors and innuendoes from which she could not extract herself; or

• Lillian Knox was a shrewd, calculating manipulator who convinced others to carry out her bidding, including perhaps murder, and lived the rest of her life in trouble with authorities or hiding from them.

Newspaper accounts, her own correspondence and other sources support both theories, but not enough to consider either conclusive.

During her 75 years, Lady Bountiful impacted in some way the lives of everyone around her— the people of East Texas, her family and friends, a legion of lawyers and judges, associates in the Texas lumber industry and the investigators who pursued her. But in 2002, few of them were alive.

Edward J. Barrett, a Chicago attorney who befriended Lillian in the late 1950s, was one of a handful of people who could remember her with clarity. But even he admits she shielded her past with aliases, undisclosed addresses, mail drops and elusive statements.

Within her own family, there is confusion, doubt and some anger about Lillian Knox. While she had nine children, including three sets of twins, she claimed only six offspring. Three children died as infants, according to family records, but some family members believe they may have been given away or left at the door of a convent, in the same fashion she did with one daughter. Another daughter committed suicide.

Lillian's oldest child, Hiram William Ambrose, took to his grave on July 12, 2000, many of the answers to the Lillian Knox mystery.

Of all the Knox children, he knew his mother best, but he was always reluctant to share his memories during the years Lillian was alive. Only when she died in 1966 did he admit his belief that Lillian had been responsible for the death of his father Hiram and the brutal slaying in 1937 of his grandmother, Mary Axford Knox, at Dallas. Hiram also said he believed Lillian had been connected with the death of Grace Eva Briggs Knox, his father's first wife, in Polk County in 1908.

Some of Lillian Knox's children, including son Hiram, were not sure if Hiram Knox fathered them. Son Hiram discovered as an adult that he was born before Lillian married his supposed father in 1910. Among the courtroom issues raised in 1923 when

Lillian was charged with Hiram's murder were her affairs with other men, principally attorney Chester B. Collins of Lufkin.

If Hiram, and perhaps some of his siblings, were sired by other men, how did Lillian's lumberman-husband, deal with the knowledge? It seems unlikely that he was unaware that Mary Lyllian, a twin born in 1918, was given to a Catholic convent in Shreveport in 1920? Did twins Grace and Lillian, who were born in March of 1913, really die on the same day in October of 1914 or where they also given away with Hiram's knowledge?

A hint of an answer lies in Lillian's comments in a letter she wrote to Edward J. Barrett February 28, 1960: "My husband (Hiram) and I had no secrets. He knew I was married to (Chester B.) Collins. He (Hiram) loved me as much as he could love any woman, but he did not want me to leave him. He extracted a promise out of me before we were married. He told (me) all he wanted to do (and) his faults, but that I would remain his wife until he agreed. He needed a wife to protect him...from women and blackmailers...I kept (it) all a secret from outsiders (but) I kept no secret from him."

Lillian's comments lead to other questions. What were Hiram's faults? What issues did she keep secret on his behalf? What secrets were susceptible to blackmail? Did she share with Hiram her extramaritial affairs with other men, such as Collins?

Another unresolved mystery is what happened to the Knox fortune inherited by Lillian Knox following Hiram's death. She left East Mayfield with an estimated million dollars in cash, oil royalties, timberland properties and other investments, but by 1930 she was broke and had turned to forgeries, mail fraud and confidence schemes.

Another Knox mystery exists at Woodlawn Cemetery in

Texarkana. While Lillian's beloved mother, Daisy E. Marshall, lies in the cemetery's most expensive family plot—an ornate marble enclosure guarded by a tall, angel-topped monument—the identification of the four members of the Knox family who died while living in Lillian Knox's shadows are missing.

Hiram was buried in the cemetery in 1922. The remains of his first wife Grace, their teenage son Willie, and Hiram's father, Colonel William Hiram (Will) Knox, were ordered removed from a Livingston cemetery by Lillian in 1924 and purportedly reburied beside Hiram in a Knox family plot adjoining the Marshall enclosure. But the Knox plot today contains only pieces of a broken tombstone with inadequate inscriptions to identify any of the four graves. A single pine tree, apparently planted after Hiram's burial in 1922, towers over the plot. Estimated to be at least eighty years old, it is the only pine tree in the cemetery—and a symbolic reminder of the family's existence and its timber empire.

There is perhaps some justice or irony in the fact that, like the four Knoxes, Lillian Marshall Knox also lies in a unmarked grave a thousand miles away.

The only existing Knox tombstone standing in Texas marks the Dallas grave of Mary Axford Knox, whose murder ended the dynasty of the family which came to Texas in 1900 to build one of the state's largest lumber fortunes—a fortune that fell, either by accident or design, into the hands of Texas' mysterious Lady Bountiful.

An eighty-year-old pine tree is the only sentinel standing above the graves of four of the five Knox family members who came from Wisconsin to Texas in 1900 to create a multi-million dollar lumber empire. The pine is in the Knox family plot in Woodlawn Cemetery in Texarkana. No tombstones identify the graves. *Photo by Bob Bowman.*

The broken partial tombstone of Hiram Knox in Woodlawn Cemetery, Texarkana. *Photo, Bob Bowman.*

Mary Knox's monument in Grove Hill Memorial Park, Dallas, is the only tombstone left of five members of the W.H. Knox family which came to Texas in 1900. *Photo, Bob Bowman.*

Acknowledgements

A number of individuals and organizations contributed immensely to the research that made this book possible, but special acknowledgments should be made to the family of Merton B. (Bud) Knox, the late son of Lillian Knox; Virginia Burns Spears, the sister-in-law of William Hiram Ambrose Knox, and her daughter, Cecilia Fry of California; Dr. Hiram Arnold, the son of Dr. Tom Arnold, the Knox family physician in 1922; Robert Cecil McDaniel of Hemphill, Sabine County historian and author; Houston attorney Gregg C. Laswell and paralegal Suzy Hodek for locating copies of Houston newspapers from the 1920s; Wanda Bobinger of the Polk County Historical Museum, Livingston, for her research on the Knox family's business interests and life in Polk County; Joyce Bant for permission to use excerpts from her history of Knox Mills, Wisconsin, *Culture and Continuity of Knox Mills, Wisconsin*; Richard S. Hollar of Fort Collins, Colorado, for his assistance in locating historical materials related to the Knox family in Wisconsin; John H. Minton, Jr., of Tyler, Texas, for his assistance with legal issues involving the Minton and Knox families; Sammie Waycasey of the Museum of Regional History Texarkana, Texas, for his research assistance with the Knox family and the lumber industry in Texarkana; and Texas State Senator Todd Staples and his staff, especially Jen Moore, for assistance with Texas state agencies.

No attorney-client privileges were breached, nor were any confidences betrayed in producing this book. Edward J. Barrett, attorney for Lillian Knox, Chicago, Illinois, 1957-1959.

BIBLIOGRAPHY

Books and Pamphlets

Armstrong, William Clinton, *The Axfords of Oxford*, New Jersey, Genealogical Researchers, Washington, New Jersey, reprinted 1984.

Blankenship, Maude Davis, *A History of My Texarkana*, Texarkana, Texas, 1950.

Block, W.T., *East Texas Mill Towns and Ghost Towns,* Volume II, The Pineywoods Foundation, Lufkin, Texas, 1995.

Bowman, Bob, Editor, *Land of the Little Angelina*, Angelina County Historical Commission, Lufkin, Texas, 1982.

Bowman, Bob, *The 35 Best Ghost Towns in East Texas*, Best of East Texas Publishers, Lufkin, Texas, 1988.

Handbook of Texas, Volumes 1 through 6, Texas State Historical Association, Austin, Texas, 1996.

East Texas Historical Journal, various editions related to the lumbering industry of Texas, East Texas Historical Association, Nacogdoches, Texas, 1985-2002.

Gomer, Robert Austin, recollections from letters written to his niece, Mrs. Helen Gomer Schluter, during his life, privately published manuscript, Hemphill, Texas, date unknown.

History of Angelina County, Texas, Lufkin Genealogical and Historical Society, Lufkin, Texas, Curtis Media Corporation, 1992.

History of Polk County, Pioneers on the Livingston-Woodville Road, Polk County Historical Survey Committee, Livingston, Texas, 1963.

Images of Texarkana, 1873-1990, Texarkana Gazette, Texarkana, Texas, 1991.

Lower Rio Grande Valley of Texas and its Builders, Profile of Lillian Knox, 1931.

Maxwell, Robert; Baker, Robert, *Sawdust Empire: The Texas Lumber Industry, 1830-1940,* Texas A&M University Press, College Station, Texas, 1983.

McDaniel, Robert Cecil, *A Pictorial History of Sabine County, Texas*, Books 1, 2, 3 and 4, Texian Press, Waco, Texas, 1994, 1996, 1998, 2000.

McDaniel, Robert Cecil, *Sabine County Texas; The First One Hundred*

Years (1836-1986), Texian Press, Waco, Texas, 1987.

Reed, St. Clair Griffin, *History of Texas Railroads,* 1981.

Speights, Virgie, *Old Timers of Sabine County, Texas*, S. Malone Printer, San Augustine, 1983.

Texarkana Pioneer Families, article on Thomas Lewis Latane Temple, Texarkana Pioneers Association, Roark Printing Company, 1961.

Texas Gulf Coast Historical and Biographical Record, Texas Gulf Historical Society, Beaumont, Texas, 1993.

Toole, Blanche Fenley; White, Edna McDaniel, *Sabine County Historical Sketches and Genealogical Records,* LaBelle Printing Company, Beaumont, 1972.

Verra, Mary E. (Hollar), *Genealogy, 1700-1994,* Axford, Bowen, Rosel, Hollar, Wood, Davenport and Varra families, date unknown.

Newspapers and Magazines

American Lumberman, "A Patriotic, Practical Woman of the South," Chicago, Illinois, June 8, 1918.

Beaumont Enterprise, Block, W.T., "Shooting Death of Lumberman Still A Mystery," Beaumont, Texas, October 2, 1999.

Beaumont Enterprise, Ramos, Ralph, "Gallows Tower Remains in Jail," Beaumont, Texas, December 10, 1972.

Crosscut, Texas Forestry Museum, Lufkin, Texas, First Quarter, 1990.

Dallas Morning News, Dallas, Texas, editions of March 13 through April 19, 1937.

Dallas Morning Times Herald, Dallas, Texas, editions of March 13 through April 1, 1937.

Dallas Morning News, Biffle, Kent, "Betting on a Dallas Mobster," Dallas, Texas, May 19, 2002.

Dallas Morning News, Griffith, Dotty, "Hotel Whitmore: Its Passage No Secret," Dallas, Texas, February 26, 1976.

Dallas Morning News, Parmley, Helen, "The End for Whitmore," Dallas Texas, November 30, 1969.

Jasper News-Boy, Derouen, Elaine, "It Was A Hard, Hard Life," Jasper, Texas, undated.

H*ouston Chronicle*, obituary of W.H. Knox, Houston, Texas, June 17, 1913.

Houston Chronicle, Houston, Texas, edition of March 13, 1937,

Houston Chronicle, various editions, Houston, Texas, November and December, 1922, and January, February, March and April, 1922.

Houston Post, "Sabine County Citizens Honor Beloved Woman," Houston, Texas, June 11, 1922.

Houston Post, Syers, Ed, "Gaines Ferry Opened Texas," date unknown, Houston, Texas.

Houston Post, Houston, Texas, various editions, November and December, 1922, and January, February, March and April, 1922.

Houston Post, Houston, Texas, editions of March 13 though March 19, 1937.

Houston Post, Houston, Texas, edition of March 16, 1937.

Houston Press, Houston, Texas, various editions, November and December, 1922, and January, February, March and April, 1923.

Houston Press, Waters, H.C., "Mrs. Knox's Own Story of Tragedy," Houston, Texas, January 5, 1923.

Jasper News-Boy, "Knox Generosity Legendary Among Sabine County Residents," author unknown, Jasper, Texas. June 9, 1999.

Kansas City Star, "The Tragic Climax in the Career of a Texas Lady Bountiful," Kansas City, Missouri, January 14, 1923.

Lufkin Daily News, Lufkin, Texas, edition of November 26, 1922.

Master Detective, Harris, Margie, "Texas' Amazing Knox Riddle," Dunellen, New Jersey, January, 1938.

Nacogdoches Daily Sentinel, Nacogdoches, Texas, edition of January 4, 1923.

New York Times, "A Grand Century, But a Modest Birthday," New York, N.Y., June 10, 2002.

Polk County Enterprise, various editions, Livingston, Texas, 1907-1902.

Redland Herald, Nacogdoches, Texas, edition of December 30, 1922.

Southern Industrial and Lumber Review, "Yellow Pine vs. White Pine," January, 1900; Lumber Notes, April, 1903.

Texarkana Gazette, Texarkana, Texas, editions of March 15 through April 24, 1937.

Almanacs and Directories

Beaumont City Directory, 1914 through 1938, Tyrell Library,

Beaumont, Texas.

Texarkana City Directory, 1889 through 1901, Texarkana Library, Texarkana, Texas.

Manuscripts, Thesis and Speeches

Bant, Joyce, *Culture and Continuity of Knox Mills, Wisconsin*, manuscript, June, 1985.

Bowman, Bob, *Lady Bountiful and the Texas Trail of Murder*, speech, East Texas Historical Association, Fall Program, Nacogdoches, Texas, September 21, 2002

Barrett, Edward J., *The Riddle of Lillian Knox,* unpublished manuscript, Chicago, Illinois, 1996.

Government Agencies

Caddo Parrish, Louisiana, death certificate, Mary Clement Knox, January 25, 1988.

Kankakee County, Illinois, death certificate, Lillian Knox, December 17, 1966.

Polk County, Texas, County Clerk and District Clerk, various documents related to W.H. Knox and Knox Lumber Company, Livingston, Texas, 1902 though 1913.

Sabine County, Texas, County Clerk and District Clerk, Texas, various documents related to Knox Lumber Company, Hemphill, Texas, 1912 through 1931.

Staples, Senator Todd, and assistant Jen Moore, Texas State Senate, Austin, Texas, 2002.

Texas Historical Commission, Austin, Texas, history of East Mayfield, Texas, 1912-1937.

Texas State Department of Health, Vital Statistics Division, Austin, Texas.

U.S. Census Bureau, Census for Sabine County, Texas, 1910 and 1920.

U.S. Department of Justice, Women's Reformatory, various documents related to Lillian Knox, Alderson, West Virginia, 1939 through 1945.

U.S. District Court, Western District of Louisiana, District Clerk, various documents related to Lillian Knox, Shreveport, Louisiana, 1939.

Texas Secretary of State, business incorporation records, Austin, Texas.

Interviews

Arnold, Dr. Hiram, Lufkin, Texas, various interviews by Bob Bowman, December, 2001 through May, 2002.

Barrett, Edward J., Chicago, Illinois, interview by Bob Bowman, May 31, 2002.

Buckley, Ivy D., tape of interview by Dwayne Buckley, Beaumont, Texas, date unknown.

Cole, Robert, St. John's, Arizona, telephone interview by Bob Bowman, June 6, 2002.

Eddings, Marguerite, Hemphill, Texas, interview by Bob Bowman, May 7, 2002.

Fry, Cecilia, Folsom, California, telephone interview by Bob Bowman, June 3, 2002.

Knox, Hiram Ambrose, video interview by Knox family, Dallas, Texas, September 4, 1991.

Knox, Merton, family members, Dorothy Knox, Mark Knox and Cindy Knox Fuller, by Bob Bowman, June 3, 2002.

McDaniel, Cecil, Hemphill, Texas interview by Bob Bowman, May 7, 2002.

McFarland, Edward, Lufkin, Texas, interview by Bob Bowman, June 19, 2002.

Minton, John H. Jr., Tyler, Texas, interview by Bob Bowman, March 25, 2002.

Minton, John W., Hemphill, Texas, interviews by Robert Maxwell, May 13, 1956, and August 5, 1958.

Nethery, Wendell, Milam, Texas, interview by Bob Bowman, May 7, 2002

O'Toole, Blanche, Hemphill, Texas, interview by Bob Bowman, May 7, 2002.

Spears, Virginia Burns, Folsom, California, telephone interview by Bob Bowman, June 5, 2002

Temple, Arthur, Jr., Diboll, Texas, interview by Bob Bowman, May 29, 2002.

Thomason, Jerome, Polk County, Texas, interview by Bob Bowman, June 26, 2002.

Upton, Robert, Polk County, Texas, interview by Bob Bowman.

June 14, 2002.

White, Wilma Placker, Polk County, Texas, interview by unknown interviewer, September 11, 1987.

Collections

Bowman, Bob, Lufkin, Texas, collection of newspaper articles, photos and other materials related to Knox Lumber Company, East Mayfield, Texas.

Knox, Lillian, Chicago, Illinois, correspondence to Edward J. Barrett and others, 1957 to 1961.

Hollar, Richard S., Colorado Springs, Colorado, correspondence and various documents related to Knox and Axford families in Wisconsin and Texas, 1999 through 2002.

Houston Public Library and Clayton Library, Houston, Texas,various documents related to the W.H. Knox family, 1922 through 1937.

Huffman Library, Hemphill, Texas, various records related to Knox Lumber Company, 1912-1924.

Knox Family Bible, in possession of Hiram Knox family, Auburn, California.

Knox photograph album, 1918-1920, in possession of Merton Knox family, Dallas, Texas.

Kurth Memorial Library, Lufkin, Texas, various documents related to the Knox, Collins and Mantooth families.

Polk County Memorial Museum and Polk County Historical Commission, Livingston, Texas, various documents related to Knox Lumber Company, 1903-1924.

Stephen F. Austin State University, Steen Library, Nacogdoches, Texas, various documents related to the Texas lumbering industry and Knox Lumber Company.

Texas Forestry Museum, Lufkin, Texas, various documents related to Knox Lumber Company.

Texarkana Museum of Regional History, various documents related to the Knox family and the lumber industry, Texarkana, Texas.

Texarkana Public Library, various documents related to the Knox family, Texarkana, Texas.

Tyrell Historical Library, Beaumont Public Library System, Beaumont, Texas, various documents related to W.H. Knox family,

Beaumont, and the Texas lumbering industry, 1920 through 1930.

Special Assistance

Bobinger, Wanda, Polk County Historical Commission, Livingston, Texas.

Laswell, Gregg C. and Hodek, Suzy, Houston, Texas for locating copies of Houston newspapers from the l920s.

Waycasey, Sammie, Museum of Regional History, for his assistance in locating the Knox and Marshall gravesites in Woodlawn Cemetery, Texarkana, Texas.

Zaragoza, Nora, Grove Hill Memorial Park, for assistance in locating Mary Knox gravesite, Dallas, Texas

Web Sources

Ancestry.Com, various entries related to W.H. Knox family, April, 2002.

TempleForest.Com, history of Temple Lumber Company and Southern Pine Lumber Company, Diboll, Texas, 1890s and early 1900s.

Toledo-Bend.Com, history of East Mayfield and Sabine County, Texas, Many, Louisiana, 2002.

Texas Forestry Museum, Lufkin, Texas, history of Knox Lumber Company and the Texas lumber industry, 1900 to 1930.

Index